THE EFFICIENT DECK HAND

A guide to the Department of Trade Examinations
for Efficient Deck Hand and Able Seaman.

A companion book to: – "Survival at Sea – The Lifeboat and Liferaft".

BY

C. H. WRIGHT

NATIONAL SEA TRAINING SCHOOL
(LIVERPOOL).
(Retired).

SAFETY DOES NOT HAPPEN
It is the reward of care, thought and good organisation.

BEWARE OF SHORT CUTS
They almost invariably reduce the safety margin.

GLASGOW
BROWN, SON & FERGUSON, Ltd., Nautical Publishers
4-10 Darnley Street

Copyright in all countries signatory to the Berne Convention
All rights reserved

Fourth Edition — 1990
Reprinted — 1991

ISBN 0 85174 558 X

©1991—BROWN, SON & FERGUSON, LTD., GLASGOW, G41 2SD
Made and Printed in Great Britain

By the same Author

"SURVIVAL AT SEA—THE LIFEBOAT AND LIFERAFT".
The most up-to-date standard text book, specially written for candidates who are attempting the Department of Trade Examination for a Certificate of Proficiency in Survival Craft, and for pre-sea training. Invaluable to all Officers, Cadets, Petty Officers, Ratings, Fishermen, Yachtsmen and the crews of floating oil-rigs, in all departments. Currently in use in countless Nautical Training Establishments throughout the English speaking world. Recommended for inclusion in all ships' libraries by The Nautical Institute.

"THE COLLISION REGULATIONS 1977 AS AMENDED 1981—FULLY EXPLAINED"
Written for both amateur and professional seamen, it contains a copy of The Final Act of the International Conference on Revision of the International Regulations for Preventing Collisions at Sea, 1972, as amended 1981 which came into force internationally on 1st June 1983. Fully illustrated with coloured diagrams suitably explained.
"Meets this vital need for guidance on the practicable application of the Regulations". *Lloyd's Gazette*. 15th July 1975.
A MUST FOR EVERY NAVIGATOR.

"SURVIVAL FOR YACHTSMEN".
Written specifically for yachtsmen, both experienced and inexperienced, who, serving on a vessel equipped with an inflatable liferaft, will want to know how it works, how to use it and what to do if ever they are so unfortunate as to be castaway.
"A masterpiece of practical information on survival afloat, in the event of a disaster, every sea-going yachtsman should study". The *Nautical Magazine*, April 1977.

"PROFICIENCY IN SURVIVAL CRAFT CERTIFICATES".
Specifically written for candidates attempting to pass Certificates of Proficiency in Survival craft. Contains all the necessary information.

Obtainable from the Publishers.
    BROWN, SON & FERGUSON, LTD.,
    4-10 DARNLEY STREET,
    GLASGOW, G41 2SD.

and from Nautical Booksellers

FOREWORD

I was interested when the Author, Captain C.H.Wright, asked me to read the manuscript of this book and I checked my diaries to find that I had examined 2,160 seamen for their Able Seaman and Efficient Deck Hand Certificates (not all passed) during my 18 years service as a Nautical Surveyor in the Board of Trade at Liverpool.

This book more than covers the syllabus and should be helpful to candidates studying for this examination and it makes a very good companion book to "Survival at Sea—The Lifeboat and Liferaft" by the same Author and now in its third edition.

I am glad that Captain Wright has mentioned Pilot Ladders and quoted the regulations for the correct rigging of the Pilot Station.

Safety and Fire have been emphasized to a praise-worthy degree; I frequently tell seamen that there is nothing cowardly about donning safety harness and lifeline, a good seaman is never foolhardy.

There are many "Nevers" in this book, all maxims based on accidents and happenings which have occurred, budding seamen will do well to take these warnings to heart. Here is another "Never"—Never, when engaged in brushing down a funnel preparatory to painting and using a longhandled deck-scrubber, insert the broom-handle down the exhaust pipe to use the broom-head as a seat. A seaman, he had sailed with me during the war, actually did this, the exhaust valve lifted and the ship blew-off, needless to say, the seaman was killed.

A word of advice to candidates appearing before the Nautical Surveyors of the Department of Trade, put on your best appearance, this is important for any oral examination or interview and I can assure you it helps a lot.

Finally, I congratulate Captain Wright on an excellent job of seamanship and hope it will ease the task of both pupils and Instructors at Nautical Training Establishments and make it more certain for D.T.I. Examiners to be able to pass candidates.

"C.G.CUTHBERTSON"

Commander, D.S.C.,R.D.,R.N.R.(Rtd).
Younger Brother of Trinity House.
Honorary Seamanship Examiner to The Indefatigable.
Marine Surveyor.

Liverpool, March 1973.

PREFACE.

In writing this text book, my aim has been to try and fill a gap by providing an up-to-date guide, for Ordinary Seamen, Carpenters, Engine-room Ratings attending an Engine to Deck Conversion course, Ex-Royal Naval Ratings joining the Merchant Service, D.H.U's, and Deck Cadets, who are attempting the Department of Trade Examination for Able Seaman or Efficient Deck Hand. I have also tried in some small measure to give in addition, details of some of the more modern and complex equipment they may be called upon to use.

It has been presumed that a candidate for an A.B., or E.D.H. Certificate (or in the case of an engine room rating, examination pass), will be in possession of a Lifeboat Efficiency Certificate. Boatwork therefore although in the syllabus, has been omitted. Any candidate who is not in possession of a Proficiency in Survival Craft Certificate, should also study in conjunction with this book, its companion book, *Proficiency in Survival Craft Certificates* in which the subject of boatwork has been more than adequately covered for candidates attempting the Proficiency in Survival Craft Certificate.

It is also hoped that many schools which provide a measure of pre-sea training for a proportion of their pupils, Sea Scouts and Sea Cadets, will find, much useful material within.

I wish to acknowledge my gratitude to all those who have so kindly assisted me in the writing and in particular:—

Commander C.G. Cuthbertson, D.S.C., R.D., R.N.R.(Rtd), Younger Brother of Trinity House, Marine Consultant.
Captain H.C. Large, Master Mariner, (a valued colleague, now retired).
Mr. H. Hambling, Miss S.J. Thorpe and other members of the B.S.F. (Liverpool)
Mrs. E. Evans and Mr. C.M. Wright, M.C.M.S.

Also the following Companies for their help with advice, literature and photographs.

Messrs.   Angus Fire Armour.
          Bibby Bros & Co.
          British Paints and Chemicals. Torpedo Marine Paints Division.
          British Ropes Ltd.
          S.G. Brown Ltd.
          Clarke Chapman-John Thompson Ltd.
          C.P. Ships.
          Cunard-Brocklebank-Atlantic Container Division.
          The DeVilbiss Company Ltd.
          J and J Denholm (Management) Ltd.
          Donkin & Co.
          Draeger Normalair Ltd.
          Ellerman Lines Ltd.
          Fyffes Group Ltd.
          Hawkins & Tipson Ropemakers Ltd.
          T & J. Harrison Ltd.

Samuel Hodge & Sons Ltd.
F.A. Hughes & Co.Ltd.
L. & G Fire Appliance Co. Ltd.
P & O Photographic Library.
MacGregor & Co (Naval Architects) Ltd.
Manchester Liners Ltd.
Marland Designs.
Martindale Electric Co.Ltd.
M.E.P. Co.Ltd.
Mine Safety Appliances Co. Ltd.
Mitchell Swire Ltd.
J. Morice & Co. Ltd.
Overseas Containers Ltd.
Respirex Ltd.
J. Sewill Ltd.
Shandon Southern Instruments Ltd.
Shell Tankers (U.K.) Ltd.
Siebe Gorman & Co. Ltd.
Silver Line Ltd.
Thomas Walker & Son Ltd.
The Welin Davit and Engineering Co. Ltd.
Whessoe Systems and Controls Ltd.
Wilson, Walton International Ltd.
Eastwood and Dickinson Ltd.

The "M" Notices are reproduced by permission of the Department of Trade.

The illustrations from the articles on Union Purchase Rigs and Rolling Hatch Covers from the October 1964 and January 1965 issues respectively of "ACCIDENTS" are reproduced with the permission of the Controller of Her Majesty's Stationery Office.

                        C.H. Wright.
                        Principal, National Sea Training Schools,
                        Liverpool. March 1973.

Note. The contents of this book should provide a good basis of general knowledge for most of the equipment likely to be encountered on a British ship. However, readers are advised that the equipment and fittings supplied to any particular ship may differ in make, description and usage, from those which are described and illustrated in this book. They should therefore always make themselves familiar with the equipment provided on any ship on which they may be serving.

CONTENTS

CHAPTER 1.	Regulations governing the issue of a Certificate as A.B. or E.D.H.	11
CHAPTER 2.	Common Nautical Terms.	16
	Time bells and watches.	19
	Emergency Signals.	20
	Lookout duties and signals.	20
	Navigation lights.	21
	Flags and signals.	22
CHAPTER 3.	The names and functions of various parts of a ship.	26
	Construction.	34
	Conventional cargo vessels.	38
	Passenger ships.	39
	Bulk carriers.	39
	LASH. (Lighter aboard ship system).	49
	Container Ships.	49
	Roll on/Roll off (Ro-Ro) ships.	49
CHAPTER 4.	COMPASSES, HELM ORDERS AND STEERING GEAR.	
	Compasses.	52
	The magnetic compass.	53
	The gyro compass.	57
	Steering gear.	57
	Helm orders.	62
CHAPTER 5.	THE PATENT LOG AND HAND LEAD LINE.	
	The patent log.	65
	The hand lead line.	67
	The deep sea hand lead line.	69
CHAPTER 6.	ANCHORS, CABLES AND WINDLASS.	
	Anchors and cables.	70
	The windlass.	73
	Anchoring.	76
	Accident prevention.	76
	"M" Notices.	
CHAPTER 7.	FIREFIGHTING AND LIFESAVING APPLIANCES.	
	Fire prevention.	77
	Fire fighting.	78
	Portable fire exrtinguishers.	79
	Major fire appliances.	81
	Foam, low, medium and high expansion.	85
	General, precautions.	85
	Fixed installations. Carbon dioxide, sprinklers and fire alarms.	86 / 87
	Fresh air breathing apparatus.	87
	Compressed air breathing appartus.	92
	Gas masks.	92
	Dust masks.	92
	Protective clothing.	92

	A gas protection suit.	93
	Protective helmets.	96
	Goggles.	96
	Gloves.	96
	Footwear.	96
	Gas tight torches.	96
	Safety belts.	96
	Atmosphere sampling equipment.	98
	Davy-lamp.	98
	Fixed Installations.	98
	Explosimeter.	98
	Fryrite gas analyser.	102
	Multi gas detector.	102
	The "Neil-Robertson" stretcher.	102
	Resuscitators.	102
	Various terms applicable to safety precautions.	103
	Carriage of dangerous goods labels.	108
	Accident prevention.	110
	"M" Notices.	111
CHAPTER 8.	ROPE AND ROPEWORK.	
	Types of rope.	125
	Natural fibres.	125
	Synthetic fibres.	125
	Preformed galvanised steel wire rope.	125
	Mixed wire and fibre ropes.	126
	Rope construction.	126
	Breaking strains and safe working loads.	126
	Care in the use of rope.	128
	Rope measurement.	129
	Types of cordage.	130
	Various ropes.	132
	Knots, bends and hitches.	134
	Riding a stay.	137
	Whippings.	140
	Seizings.	143
	Stoppers.	143
	Ladders.	145
	Pilot ladders.	145
	Stages.	149
	Accommodation ladder.	151
	Moorings.	154
	Accident prevention.	158
	"M" Notices.	159
CHAPTER 9.	SPLICING.	
	Splices.	161
	Back splice.	161
	To splice any hawser laid rope.	162
	Eye splice.	164
	Cut splice.	166
	Short splice.	166
	To splice squareline or multiplait rope.	167
	To splice flexible steel wire rope (six stranded).	172
	Serving.	175

CHAPTER 10.	LIFTING TACKLE, **BLOCKS, PURCHASES,** MASTS, DERRICKS, WINCHES.

BLOCKS.
Parts of a block. 178
Types of blocks. 179
Care of blocks. 183
Purchases. 187
Types of purchases or tackles. 187
Chain hoists. 188
Masts. 190
Topmasts. 191
To lower a telescopic topmast. 192
DERRICKS.
To top a derrick. 195
Working cargo with conventional derricks. .. 197
Care of derricks and cargo gear. 199
Jumbo derricks. 200
Cranes. 202
Winches. 206
Electric winches. 208
Hydraulic winches. 208
Self-tensioning winches. 208
Accident prevention. 208

CHAPTER 11. HATCHES, HOLDS AND DEEP TANKS.
Hatches. 209
To strip a conventional hatch. 212
To open a deep tank lid. 214
MacGregor steel hatch covers. 215
Routine of opening and closing MacGregor
Rolling hatch covers. 216
Holds. 230
Ventilation. 234
Accident prevention. 235
"M" Notices. 236

CHAPTER 12. MAINTENANCE.
Painting. 239
Various types of paint. 240
Paint application. 242
Paint brushes. 242
Rollers. 243
Conventional spray. 243
Airless spray. 243
Tools. 246
Oil lamps. 247
Rounds. 247
Hygiene. 247
Wire ropes. 247
Cleaning brass. 248
Canvas. 248
Petty pilferage. 248
Accident prevention. 248

CHAPTER 13. HAZARDOUS CARGOES.
Ammunition. 250
Corrosive and/or toxic fluids. 250
Oil, Liquid Gas and Chemicals. 250
Summary. 271
"M" Notices. 272

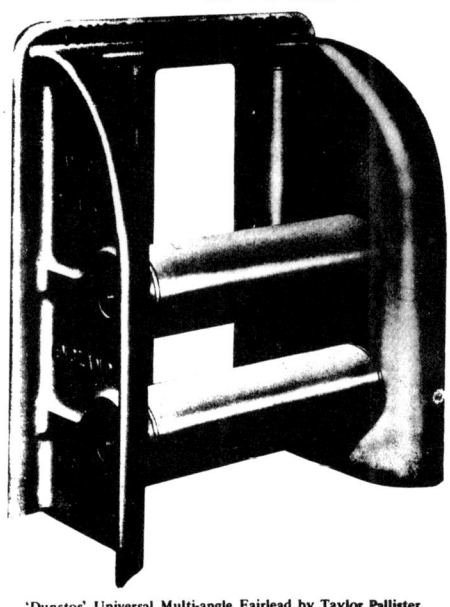

'Dunstos' Universal Multi-angle Fairlead by Taylor Pallister.
Approved for use in the St. Lawrence Seaway

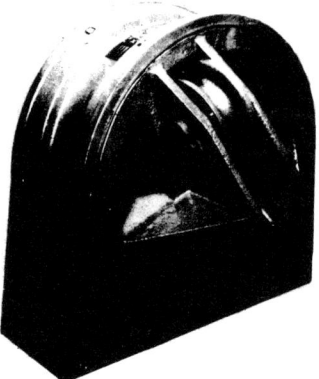

Atlas Fairleads—as used by Manchester Lines in the St. Lawrence Seaway

The fairleads comprise essentially a rotating element with 2 rope rolls in a sturdy-built casing. These rolls are fitted between 2 structural parts of the rotating element which have projecting center parts effecting that the rotating element is automatically turned into the direction of the rope. The rotating element is carried on ball races arranged at each side of the casing, the balls running on a hardened surface. Sealing of the rotating element against dirt and seawater from the outside is effected by means of a special packing. The two ball races are greased by means of lubrication nipples arranged at the outside of the casing. The rope rolls carried in bronze bushes are lubricated through the axles.

PANAMA LEAD

CHAPTER 1.
(Containing extracts from Merchant Shipping Notice No. M780).
REGULATIONS GOVERNING THE ISSUE OF A CERTIFICATE OF COMPETENCY AS AB.

A Certificate of Competency as AB is issued under the Merchant Shipping Act of 1948 and Merchant Shipping (Certificates of Competency as AB.) Regulations 1970. No seaman engaged in a British ship registered in the United Kingdom may be entered on the articles as A.B., unless he is the holder of a certificate.

A certificate may be issued without examination on payment of a fee to any seaman who was serving on articles on a British ship as Certified Deck Hand, AB, or in a superior deck capacity prior to 1st May 1952, Application should be made on Form EXN 50B at any Mercantile Marine Office. No certificate can be issued to any seaman not exempted from examination by this clause, until the seaman has:—
(a) Passed the qualifying examination.
(b) Attained the age of 18 years.
(c) i. Served for 3 years (less remission if any) as a deck rating.
   ii. Served for 3 years 6 months (less remission if any) as a GP rating. (Deck and GP service may be mixed and counted proportionately).
(d) Is in possession of a Certificate of Efficiency as a Lifeboatman or a Certificate of Proficiency in Survival Craft.
(e) Holds a steering certificate for a period of not less than ten hours, excluding time spent under instruction.

Where these requirements are satisfied, application for a Certificate of Competency as AB, may be made at any Mercantile Marine Office on Form EXN 50A. A fee will be charged for the certificate.

A seaman of 18 years and over may apply to be examined in the qualifying examination, provided he has:—
(a) 12 months service (less remission if any) as a deck rating and is in possession of a steering certificate for not less than ten hours.
or
(b) 18 months service (less remission if any) as a GP rating and is in possession of a steering certificate for not less than 10 hours.

Not less than 25 per cent of the qualifying sea service must have been performed in ships (other than fishing boats) of 100 gross tons or more, or in sailing ships of 40 gross tons or more. The remainder of the service may be performed in any type of vessel of 15 gross tons or more but service in the smaller vessels is allowed to count at half rate only. (Qualifying service includes service as a deck rating performed on sea-going vessels of the Royal Navy).

A registered seaman intending to take the examination, should first apply to a Merchant Navy Establishment Office or to the Company by whom he is employed, who will make arrangements for the seaman to attend a course at one of the National Sea Training Schools situated in the principal ports of the United Kingdom. The seaman should then make application to take the examination at any Mercantile Marine Office on Form EXN 50C for an Efficient Deck Hand Certificate, or Form EXN 50A for an AB Certificate. The seaman will be required to produce proof of age and sea service together with a steering certificate. (In the case of application for an **AB Certificate, a Certificate of Efficiency as a Lifeboatman** must also be produced). A fee will be charged for the examination. Arrangements for the seaman to take the examination will be made by the school. The examiner will notify the candidate of the results in writing and a successful candidate must produce his Discharge Book (Dis.A.) to the Superintendent of a Mercantile Marine Office for endorsement and issue of the certificate. A successful candidate who does not qualify for an AB Certificate will be issued with an Efficient Deck Hand (E.D.H.) Certificate.

A seaman who is unable to attend a course will be directed by the Mercantile Marine Office Superintendent to the Examiner, who will make arrangements to hold the examination. In such cases as long a notice as possible should be given by the candidate, in order that the examiner may have time to make the arrangements.

Although not compulsory for them, all deck cadets, midshipmen, and apprentices are strongly recommended to obtain an Efficient Deck Hand or AB Certificate. Application for a course should be made to their Company or to the Principal of a Marine Technical College they are attending.

Carpenters are advised that time served on articles as a Ship's Carpenter is allowed to count as sea service for a deck rating.

Candidates for an Efficient Deck Hand Certificate, should whenever possible, obtain at the same time a Certificate of Efficiency as a Lifeboatman, or a Certificate of Proficiency in Survival Craft, in order that they may qualify for an AB Certificate immediately they have gained the necessary sea service.

Under certain circumstances the examination may be taken by candidates under the age of 18 years and prior to going to sea, or by an Engine Room rating without deck service who has attended an approved Engine Room/Deck Conversion Course. The result of the examination will then be entered in their Discharge Book (Dis.A) but the Certificate will not be issued until the candidate has:—
(a) Reached the age of 18 years.
(b) Obtained a steering certificate for not less than 10 hours.
(c) Served the requisite period of sea service (less remission).

The examinations are conducted by Department of Trade Nautical Surveyors at the following ports.

Aberdeen, Belfast, Blyth, Bristol, Cardiff, Falmouth, Glasgow, Great Yarmouth, Grimsby, Hull, Leith, Liverpool, London, Middlesbrough, Newcastle, Southampton, Sunderland and Swansea.

AB QUALIFYING EXAMINATION SYLLABUS
Nautical Knowledge.
1. The meaning of common nautical terms.
2. The names and functions of various parts of the ship; for example, decks, compartments, ballast tanks, bilges, air pipes, strum boxes.
3. Knowledge of the compass card $0°$ to $360°$. Ability to report the approximate bearing of an object in degrees or points on the bow.
4. Reading streaming and handing a patent log.
5. Markings on a hand lead line, taking a cast of the hand lead and correctly reporting the sounding obtained.
6. Markings of the anchor cable.
7. Understanding helm orders.
8. The use of firefighting and lifesaving appliances.

Practical Work (Tested as far as possible by practical demonstration).
9. Knots, hitches and bends in common use:—
    Reef knot.   Rolling hitch.
    Timber hitch.   Figure of eight.
    Clove hitch.   Wall and Crown.
    Bowline and bowline on the bight.
    Sheet bend, double and single
    Sheepshank.
    Round turn and two half hitches.
    Marline spike hitch.
    To whip a rope's end using plain or palm and needle whipping.
    To put a seizing on a rope and wire.
    To put a stopper on a rope or wire hawser and derrick topping lift.
10. Splicing plaited and multi-strand manila and synthetic fibre rope, eye splice, short splice and back splice.
    Splicing wire rope, eye splicing using a locking tuck.
    Care in use of rope and wire.
11. Slinging a stage, rigging a bosun's chair and pilot ladder.
12. Rigging a derrick, driving a winch, general precautions to be taken before and during the operation of a winch whether used for working cargo or for warping.
13. The use and operation of a windlass in anchor work and in warping.
    Safe handling of moorings with particular reference to synthetic fibre ropes and self tensioning winches. Precautions to be taken in the stowage of chain cable and securing the anchors for sea.
14. A knowledge of the gear used in cargo work and an understanding of its uses. General maintenance with particular reference to wires, blocks and shackles.
15. The safe handling of hatch covers including mechanical hatch covers, battening down and securing hatches and tank lids.
16. If a lifeboatman's certificate or a Certificate of Proficiency in Survival Craft is not held, a candidate will be required to satisfy the examiner that:
    (a) He understands the general principals of boat management and can carry out orders relating to lifeboat launching and operation and the handling of a boat under sail.
    (b) He is familiar with a lifeboat and its equipment and the starting and running of the engines of a power boat.
    (c) He is familiar with the various methods of launching liferafts and precautions to be taken before and during launching, methods of boarding and survival procedure.

Candidates may be of either sex and will not be required to take a written examination but will be examined orally and practically. There are no restrictions on the number of attempts that may be made, or on the frequency of the attempts.

AB Certificates granted in the following countries have the same force as those granted in the United Kingdom:—
Barbados, Canada, Ghana, Republic of Ireland, Mauritius, New Zealand, Nigeria Trinidad and Tobago, Gilbert and Ellis Islands and Malta.

E.D.H. Certificates issued after the 1st October, 1957 in the Republic of Ireland, Ghana, Nigeria, Pakistan, New Zealand, Gilbert, Ellis Islands and Malta, are acceptable as proof that the holder has passed the qualifying Examination.

REMISSION OF SEA SERVICE FOR PRE-SEA TRAINING.
NAUTICAL TRAINING SCHOOLS

Location	School	Remission
Aberdeen.	Robert Gordon's Institute of Technology	6 months
Belfast.	City of Belfast College of Technology.	6 months
Blyth.	Wellesley Nautical School.	6 months
Bristol.	Incorporated National Nautical School.	6 months
Buckie.	Buckie High School.	3 months
Cardiff.	Reardon Smith Nautical College.	6 months
	National Sea Training School.	3 months
Dover.	Prince of Wales Sea Training School.	6 months
Dundee.	Dundee Technical College.	6 months
Fleetwood.	The Nautical College. (Cadet Course)	6 months
	The Nautical College. (Junior course)	3 months
Glasgow.	University of Strathclyde.	6 months
Gordonstoun.	Gordonstoun School (Nautical Department)	6 months
Gravesend.	National Sea Training School.	3 months
Greenock.	Watt Memorial College.	6 months
Hull.	Kingston-upon-Hull School of Nautical Training	6 months
	Kingston-upon Hull Nautical College	6 months
	Trinity House Navigation Schools	5 months
Leith.	Leith Nautical College.	6 months
	Leith Nautical College (TS Dolphin).	3 months
Liverpool.	Riversdale Technical College.	6 months
	The Indefatigable & National Sea Training School for Boys.	6 months
London.	The Merchant Navy College.	6 months
	London Nautical School.	6 months
Plymouth.	Plymouth College of Technology.	6 months
Southampton.	School of Navigation, University of Southampton, Warsash.	6 months
South Shields.	South Shields Marine and Technical College.	6 months
	National Sea Training School	3 months
Stornoway	Lews Castle College. (Cadet Course)	6 months
	Lews Castle College. (Junior Course)	3 months
Yarmouth I.O.W.	St. Swithin's Nautical School.	6 months

TRAINING SCHOOLS

Remission available only to candidates who have attended a course of pre-sea training at one of the Nautical Schools listed above.

Location	School	Remission
Barrow in Furness.	Holker County Secondary School.	6 months
Dovercourt.	Harwich School.	6 months
Falmouth.	Falmouth Technical College.	6 months
Grimsby.	Grimsby College of Further Education.	6 months
Lowestoft.	Lowestoft Technical College.	6 months
Morpeth.	Amble County Secondary School.	6 months
Newcastle-on-Tyne.	Blakelaw School.	6 months
Sunderland.	Hylton Red House Comprehensive School.	6 months
Whitehaven.	Kells Secondary School.	6 months

ESTABLISHMENTS OR INSTITUTIONS.
Name.   Course.
The Sea Cadet Corps.   Cadet Petty Officers.   3 months
APPROVED ENGINE ROOM/DECK CONVERSION COURSES.
Liverpool.   National Sea Training School.   4 weeks.
South Shields.   National Sea Training School.   4 weeks.
    The maximum remission allowed for an E.D.H. Certificate is 4 weeks and the maximum remission allowed for an AB Certificate is 6 months.
    An engine room rating who has attended an approved conversion course may also count his service as an engine room rating at the rate of one half of that service up to a maximum of 6 months, this remission will count equivalent to deck rating service at sea for an E.D.H. or AB Certificate as follows, provided a conversion course has been attended. (M780 para 20.)

12 months engine room service	=	6 months		deck service.	
3 week conversion course	=		21 days	,,	,,
8 months general purpose service.	=	5 months	10 days	,,	,,
Total	=	12 months		,,	,,

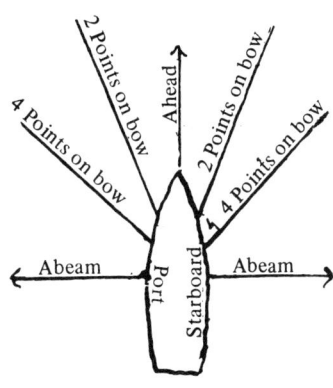

Showing method of reporting a

bearing on the bow in points.

CHAPTER 2. COMMON NAUTICAL TERMS.

A.B.	—	Able Seaman.
Abaft.	—	Behind.
Abeam.	—	At right angles to the fore and aft line of the ship.
Aboard.	—	On the ship.
Adrift.	—	Loose/Afloat without any means of propulsion, at the mercy of the wind and sea.
Afloat.	—	Laying on top of the water.
Aft.	—	Towards the stern.
Amidships.	—	In the centre line of the ship.
Articles.	—	An Agreement between the ship's Master and the crew.
Astern.	—	Behind the ship.
Avast.	—	Stop hauling on a rope.
Aweigh.	—	Said of an anchor when it is broken out of the ground.
Awning.	—	Canvas or plastic roof erected as protection from the sun.
Ballast.	—	Water or other weight carried in a ship without profit.
Beam.	—	Distance across the widest part of a ship.
Becket.	—	A loop of rope.
Below.	—	Below deck.
Bend on.	—	Tie one rope to another.
Blue Peter.	—	The International Code Flag P.
Bo'sun.	—	Deck foreman.
Bowse-in.	—	Bind in tightly.
Breaker.	—	A small cask used to carry fresh water. A wave with a broken crest.
Burlap.	—	Sackcloth.
Canvas.	—	Material used to make sails, tarpaulins, boatcovers, etc.
Capsize.	—	Turn over.
Cargo.	—	Goods carried for profit.
Carry away.	—	For a rope to break.
Cat walk.	—	Temporary gangway laid on top of deck cargo.
Chippy.	—	Carpenter.
Clear.	—	Keep a rope free of kinks and tangles.
Cock.	—	Valve to control the flow of liquid along a pipe.
Coil.	—	Stow a rope in a circular form.
Come up.	—	Order to stop pulling on a rope and ease the strain so that the rope can be secured.
Crack.	—	To open a valve slightly.
Crew.	—	Personnel forming the ship's company on board.
Course.	—	The direction in which the ship is being steered.
D.H.U.	—	Deck Hand, Uncertificated.
Doctor.	—	Ship's cook.
Donkeyman.	—	Engine room crew foreman.
Draught.	—	Depth of water required by the ship to float her.
Dual Purpose Rating	—	A person qualified as both an engine room hand and A.B. or E.D.H.
Dunnage.	—	Loose wood and mats placed under or over cargo.
E.D.H.	—	Efficient Deck Hand.
Fathom.	—	Measurement of six feet (1.83m) in length.
Fid.	—	Wood spike used when splicing fibre rope— Heavy steel pin used to support a telescopic topmast.
Fleet a tackle.	—	Stretch a purchase to its full length.

Fo'castle.	—	Crew's quarters.
Forward.	—	Towards the bow.
Galley.	—	Kitchen.
Gangway.	—	Portable bridge between the ship and a quay.
Granny.	—	An incorrectly made reef knot which is unsafe.
Greaser.	—	An engine room hand.
Hands.	—	Members of the ship's crew.
Haul.	—	Pull upon.
Helm.	—	Wheel by means of which the ship is steered.
In ballast.	—	An empty ship.
Inboard.	—	Inside the hull.
Inglefield clip.	—	A patent clip mainly used for clipping flags to halyards.
Iron mike.	—	Automatic steering gear.
Jacob's ladder.	—	A thin rope side ladder.
J.O.S.	—	Junior Ordinary Seaman.
Knot.	—	Speed of one nautical mile per hour. A joining of two ropes.
Labour.	—	Roll and pitch.
Launch.	—	Small motor boat. Place a small boat in the water.
Lay.	—	The twist of the strands of a rope.
Lee Side	—	That side of the ship the wind is blowing away from.
Leeward	—	Away from the wind.
Let go.	—	Untie or free a rope.
Light ship.	—	Without cargo.
Light vessel	—	A navigational aid.
List.	—	The amount by which a ship is leaning over towards one side.
Loaded ship.	—	Full of cargo.
Log.	—	A diary of the happenings aboard a ship. A device for recording the distance a ship has travelled through the water.
L.G.C.	—	Liquid Gas Carrier.
L.P.G.	—	Liquified petroleum gas, e.g., propane, butane, etc.
L.N.G.	—	Liquified natural gas, e.g., methane.
Lubber line.	—	Vertical line on the inside of a compass bowl, where it can be seen, and which is in line with the fore and aft line of the ship.
Lull.	—	Temporary easing of the wind force.
Manifest.	—	An inventory of possessions, stores and cargo, required by the Customs Authorities.
Make fast.	—	Secure.
Make water.	—	To take in water through a leak.
Man.	—	To provide with men.
Master.	—	Legal title for a ship's captain.
Mate.	—	Deck Officer (old and honourable title).
Mess.	—	Dining room.
Moor.	—	To secure a ship alongside a quay or with two anchors.
Muster.	—	Assemble in a particular place.
Munday.	—	Heavy hammer.
Navigate.	—	Proceed from one place to another.
N.U.C.	—	Not under command.

Oakum.	— Rope that has been unlaid, the yarns teased out and impregnated with stockholm tar.
Outboard.	— On the outside of the ship or towards the outside.
Overboard.	— In the water.
Pallet.	— A tray carrying cargo. May remain with the cargo for easy discharge.
Paravane.	— A device towed from the bow to cut moored mines adrift.
Part.	— When a rope breaks it is said to "part".
Pay off.	— Discharge a crew at the end of a voyage.
Peggy.	— Crew Mess Steward—So named because this work was performed by a seaman who had lost a leg in action and wore a wooden leg. (Peg Leg).
Portside.	— Left hand side of a ship, facing forward.
Quarter.	— After part of the ship.
Quartermaster.	— Man employed as a helmsman.
Reeve.	— Pass the end of a rope through an opening.
Ringbolt.	— A secured bolt that has a ring attached to it.
Saloon.	— Officers' dining room.
Scotchman.	— Any sleeve placed on rigging to take chafe.
Scull.	— Propel an open boat by means of an oar shipped at the stern
Sea time.	— Service rendered as a member of the crew of a ship on articles.
Set.	— Direction in which a current flows. A cold chisel.
Shackle.	— Two piece "D" shaped connector, made of steel.
Sharks mouth.	— Slit in each end of a boat cover, through which the falls are led to the lifting hooks.
Ship.	— Put something in its working position, or take it aboard. A large vessel.
Side ladder.	— Rope ladder used for putting over the ship's side.
Sign on.	— Join a ship for a voyage as a member of the crew.
Sign off.	— Cease to remain a member of the ship's crew.
Skipper.	— Legal title for person in charge of a fishing vessel.
Slip.	— Leg go of.
Span.	— Length of rope with an eye each end, stretched between two points
Spreaders.	— Short pieces of wood fitted between the gunwale and strongback, to help support a boat cover.
Squeegee.	— Rubber strip attached to wood, used as a broom for removing excess water from wood decks.
Stability.	— Righting force of a ship.
Starboard side.	— Right hand side of a ship, facing forward.
Steer.	— Keep the ship heading in a required direction.
Stern.	— Rear end of a ship.
Stove in.	— Anything broken by bad weather is said to be "stove in".
Stow.	— Put an object neatly in its proper place.
Strongback.	— A fore and aft beam over an open boat that supports a boat cover, when the boat is in its stowed position.
Sugi	— Water mixed with soda, used to clean paintwork.
Surf.	— Broken sea caused by waves breaking on the shore.

Term	Definition
Survey.	– Periodical inspection of a ship and her equipment.
Sweat down	– Tighten a rope.
Thwartships.	– Across the ship from one side to the other.
Topside.	– Paint used on a ship's side above the loaded waterline. On deck.
Tow.	– Pull a vessel through the water by means of a rope.
Transverse.	– Thwartships.
Trice.	– To bind in.
Trim.	– Difference in the amounts of water drawn by the bow and stern of a ship
Turn to.	– Start work.
Turn up.	– Make secure a rope on the bitts.
Trough.	– Hollow between two waves.
Under way.	– Proceeding through the water.
Unship.	– Remove an object from its working position.
Wake.	– Disturbed water left astern of a moving ship.
Warp.	– Move a ship by means of ropes. A mooring line.
Wash down.	– Hose the decks down.
Watch.	– Period of duty.–Men on duty.
Waterborne.	– Afloat.
Way.	– Impetus of a ship through the water.
Weather side.	– The side of a ship that has the wind blowing on to it.
White horses.	– Fast running waves with white crests.
Windward.	– Towards the wind.

Note: A ship "Goes Ahead", "Stops" or "Moves Astern", "Drifts" to leeward and is "Set" by a tide or current.

TIME BELLS AND WATCHES.

Middle watch.	Morning watch.	Forenoon watch.
0030 hrs. – 1 bell.	0430 hrs. – 1 bell.	0830 hrs. – 1 bell.
0100 hrs. – 2 bells.	0500 hrs. – 2 bells.	0900 hrs. – 2 bells.
0130 hrs. – 3 bells.	0530 hrs. – 3 bells.	0930 hrs. – 3 bells.
0200 hrs. – 4 bells.	0600 hrs. – 4 bells.	1000 hrs. – 4 bells.
0230 hrs. – 5 bells.	0630 hrs. – 5 bells.	1030 hrs. – 5 bells.
0300 hrs. – 6 bells.	0700 hrs. – 6 bells.	1100 hrs. – 6 bells.
0330 hrs. – 7 bells.	0720 hrs. – 7 bells.	1120 hrs. – 7 bells.
0345 hrs. – 1 bell.	0745 hrs. – 1 bell.	1145 hrs. - 1 bell.
0400 hrs. – 8 bells.	0800 hrs. – 8 bells.	1200 hrs. – 8 bells.

Afternoon watch.	1st Dog watch.	First watch.
1230 hrs. – 1 bell.	1630 hrs. – 1 bell	2030 hrs. – 1 bell.
1300 hrs. – 2 bells.	1700 hrs. – 2 bells.	2100 hrs. – 2 bells.
1330 hrs. – 3 bells.	1730 hrs. – 3 bells.	2130 hrs. – 3 bells.
1400 hrs. – 4 bells.	1800 hrs. – 4 bells.	2200 hrs. – 4 bells.
1430 hrs. – 5 bells.		2230 hrs. – 5 bells.
1500 hrs. – 6 bells.	2nd Dog watch.	2300 hrs. – 6 bells.
1530 hrs. – 7 bells.		2330 hrs. – 7 bells.
1545 hrs. – 1 bell.	1830 hrs. – 1 bell.	2345 hrs. – 1 bell.
1600 hrs. – 8 bells.	1900 hrs. – 2 bells.	2400 Mid – 8 bells.
	1930 hrs. – 3 bells.	
	1945 hrs. – 1 bell.	
	2000 hrs. – 8 bells.	

N.B. Six bells are never rung at 7.00 p.m. This was the signal for the mutiny at the Nore.

EMERGENCY SIGNALS.
Emergency Stations.
Not less than seven short blasts followed by a prolonged blast on the ship's whistle or siren.
A continuous ringing of electric bells and/or gongs, on ships provided with them.

Fire Stations.
A continuous ringing of the ship's bell or a gong.
When either of the above signals are sounded each member of the crew will put on warm clothes and a lifejacket. Then proceed to his emergency station and carry out the duties assigned to him on the Emergency Muster List. He will then stand-by for orders.

LOOKOUT DUTIES AND SIGNALS.
Every ship is required to post a look-out man from sunset to sunrise and at all times during bad visibility. Normally the look-out man will be stationed either on the fo'castle head or in the crow's nest but he may be on the bridge, especially in bad weather.

The duty of the look-out man is to keep a sharp look-out from right ahead to the beam on either side of the ship and report all lights and objects that he sights. In the absence of a telephone, the following signals will be used:—
One stroke on the bell. Light or object on the starboard bow.
Two strokes on the bell. Light or object on the port bow.
Three strokes on the bell. Light or object dead ahead.
He must also be able to give the approximate bearing in points, report near or far, and name the object sighted if required.

The look-out man will also repeat the time bells, when these are struck on the bridge, and report the navigation lights burning brightly (or otherwise) on the hour, every hour.

Under no circumstances will the look-out man leave his post, without being properly relieved.

The stern light is to be reported as burning (or otherwise) every four hours, normally this will be done by the man who goes aft to read the patent log at the end of each watch. If there is no log streamed, it is to be reported by one of the watch-keepers before going off watch.

When the ship is at anchor, the look-out man will normally keep an anchor watch on the fo'castle head. He should keep a close watch on the anchor cable and if the cable is heard, seen or felt (by placing a hand on the cable) to be vibrating, the Officer of the Watch is to be immediately informed that the ship may be dragging her anchor.

At anchor in fog, the look-out man will ring the ship's bell rapidly for five seconds every minute. On ships over 350 feet (106.5m) in length, a second man will be stationed aft to sound a gong rapidly for five seconds every minute.

NAVIGATION LIGHTS.

A powered vessel under way is required to carry five navigation lights.

A white mast head light 40 feet (12m) (when the beam of the ship if 30 feet or more) above the hull. This light is to be visible from right ahead to two points abaft the beam on either side. The light is normally screened underneath to prevent glare at deck level. Vessels over 165 feet (50m) in length are also required to carry a second white mast head light of similar construction at least 15 feet (4.5m) higher and astern of the first light. These two lights are to be placed in line with the keel and are to be visible six miles on a clear night.

On the port side a red light, visible from right ahead to two points abaft the beam on the port side. On the starboard side a green light, visible from right ahead to two points abaft the beam on the starboard side. These two lights are to be screened so that the port light cannot be seen from starboard or the starboard light from port. They are to be visible three miles on a clear night.

At the after end of the vessel, a white light visible for 6 points from right aft on each side, which shall be visible on a clear night for a distance of three miles.

Vessels being towed and vessels not under command will carry the side and stern but not the mast head lights.

Various extra masthead lights are required to be carried by vessels when towing, when hampered by picking up or laying cables or when on pilot duty. Separate lights are also required for vessels engaged in fishing, minesweeping, dredging and sailing vessels.

A vessel which for some reason is not under command will dowse her mast head lights and substitute two all round red lights one above the other, at least 6.5 feet (2m) apart, placed in a position where they can best be seen, visible for a distance of at least three miles on a clear night. By day she shall carry two black balls one above the other and at least 6.5 feet (2m) apart, where they can best be seen. These lights and shapes are NOT signals of distress.

A vessel at anchor is required to carry one all round white light, in the fore part of the vessel where it can best be seen. Vessels of 165 feet (50m) in length and over shall carry the light not less than 20 feet (6m) above the hull and shall also carry a second all round white light at the after end of the vessel, not less than 14.75 feet (4.5m) lower than the forward light. These lights are to be visible for a distance of three miles on a clear night. By day they shall carry in the fore part of the vessel, one black ball.

A vessel aground in or near a fairway is required to carry both anchor lights and not under command lights. By day she shall carry three black balls in a verticle line, not less than 6.5 feet (2m) apart, where they can best be seen.

The black balls or shapes referred to above are required to be two feet (.6m) in diameter. Cork fenders are not a suitable substitute and are not to be used as such.

Emergency oil navigation lights are carried in ships not fitted with dual electric systems and are for use, if for any reason one or more of the navigation lights should fail. All ships carry oil not under command and anchor lanterns for emergency use.

Navigation lights are to be exhibited at all times between sunset and sunrise and at any other time when it may be deemed necessary.

FLAGS AND SIGNALS.

A knowledge of the International Code of Signals is not required by a candidate for the E.D.H. or A.B. examination. However some general knowledge of the more common flag hoists and signals may be expected.

Flag flown from the jackstaff.
For decorative purposes only, either the ship's house flag or a pilot jack.

Flag. flown from the foremast head. Courtesey flag.
This is the ensign of the country the ship is visiting and is flown as a mark of respect.

Signals flown from the foremast yardarm or the jumper stay.
Flag B. (a plain red burgee) At night—a red light.—I am loading or discharging dangerous or explosive cargo.
Flag G. (yellow and blue vertical stripes) ⎫
Pilot jack. (Union jack with a white border) ⎬ I require a pilot
At night—a blue flare. ⎭
Flag H. (half white-half red divided vertically)—I have a pilot on board.
Flag P. (blue with a square white centre)—I am sailing within 24 hours.
Flag Q. (all yellow)—Flown when entering port is a request for the Port Health Authority to visit and give a clean bill of health to the ship.
Ensign with a weft (knot in the flag).—I require Customs.
Ship's numbers. A four flag hoist indicating the ship's name.
Flag flown from the main mast head. House flag. This is the Company's own flag and is often raised above the mast head by means of a staff.

Flag flown from the flagpole at the stern. The ship's ensign. When at sea, the ensign may be flown from a gaff.

The ensign is hoisted at 0800 hours and lowered at sunset, it is to be hoisted flying and must not be hauled to the truck and broken out. The house flag, which may be broken out, is hoisted and lowered with the ensign, likewise the courtesey flag.
International signals may be flown at any time in daylight.

Dipping the ensign.
When passing a warship of any nationality at sea, it is customary to hoist the ensign. As the warship comes abeam the ensign is lowered. The warship will then lower her own ensign and after a pause, raise it. When the warship's ensign has been raised, the merchant ship should raise her own ensign again.

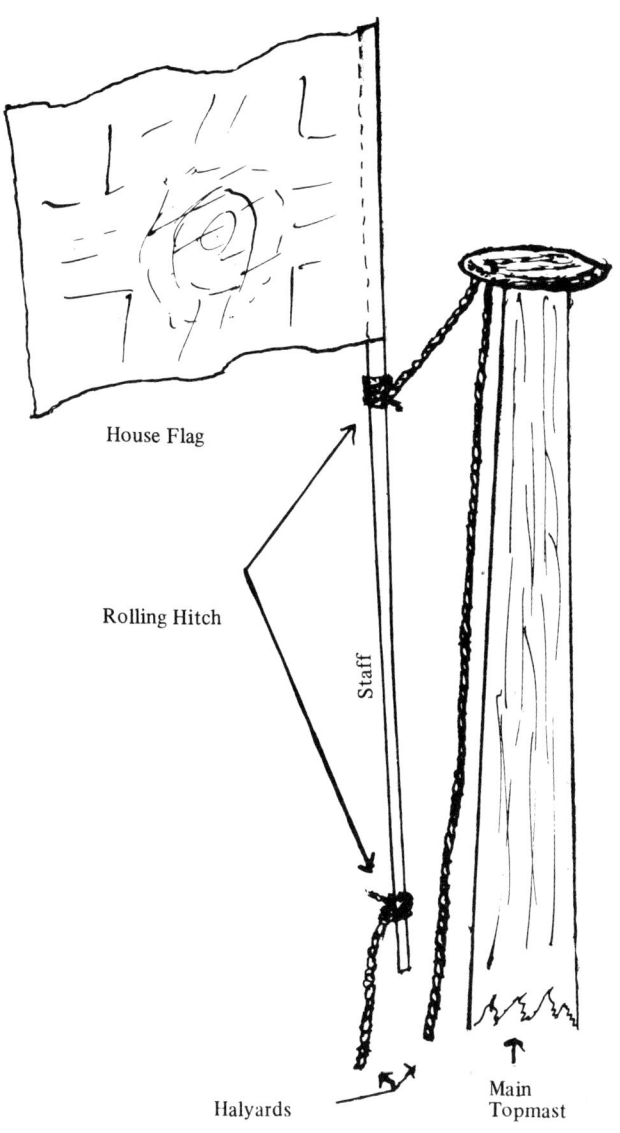

Bending on a House Flag

SILHOUETTES

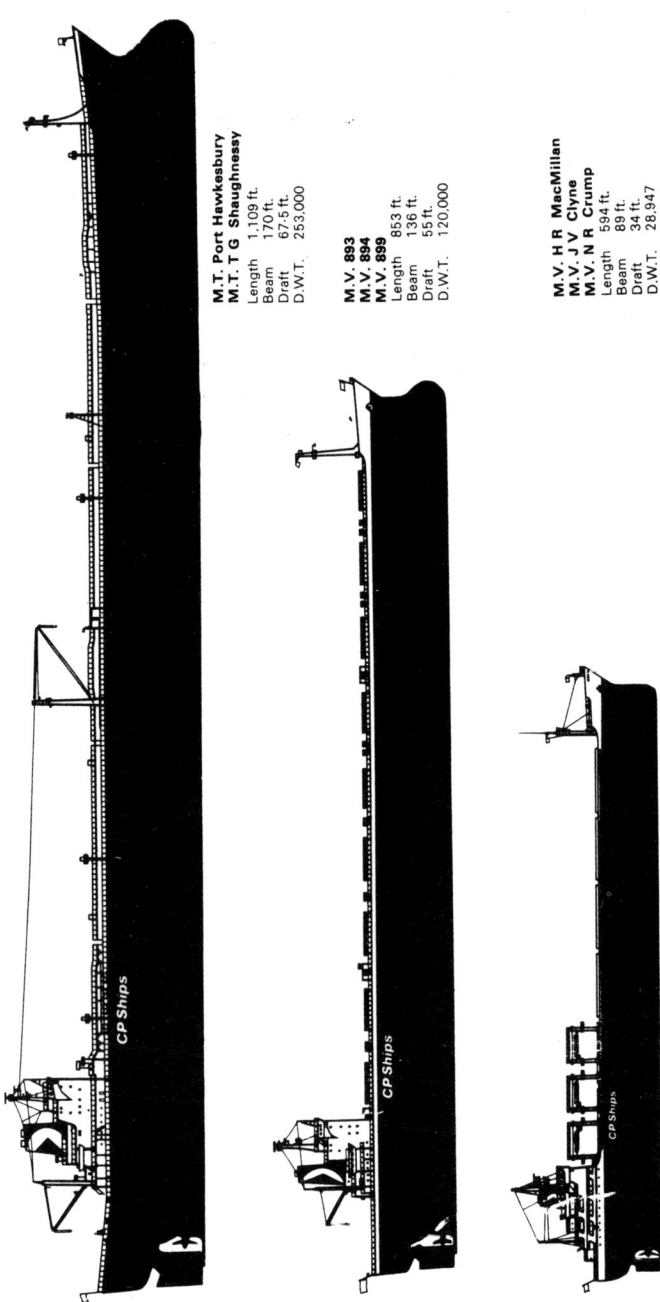

M.T. Port Hawkesbury
M.T. T G Shaughnessy
Length 1,109 ft.
Beam 170 ft.
Draft 67.5 ft.
D.W.T. 253,000

M.V. 893
M.V. 894
M.V. 899
Length 853 ft.
Beam 136 ft.
Draft 55 ft.
D.W.T. 120,000

M.V. H R MacMillan
M.V. J V Clyne
M.V. N R Crump
Length 594 ft.
Beam 89 ft.
Draft 34 ft.
D.W.T. 28,947

M.V. CP Voyageur
M.V. CP Trader
M.V. CP Discoverer
Length 550 ft.
Beam 84 ft.
Draft 30 ft.
D.W.T. 16,000

M.V. Pacific Logger
Length 486 ft.
Beam 69 ft.
Draft 29 ft.
D.W.T. 15,900

M.V. Beaverpine
M.V. Beaverelm*
M.V. Beaverfir*
Length 371 ft.
Beam 52 ft.
Draft 26 ft.
D.W.T. 6,000

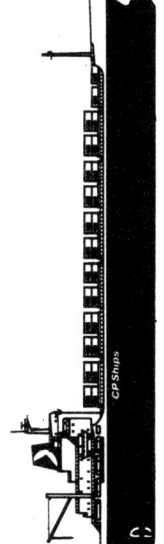

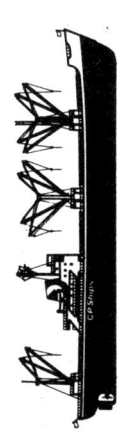

CHAPTER 3.
THE NAMES AND FUNCTIONS OF VARIOUS PARTS OF A SHIP.

Accommodation-ladder.	— A portable staircase put overside when the ship is at anchor, for the use of persons embarking and disembarking.
Aerial.	— Single wire between mastheads for use with the ship's wireless equipment.
After peak.	— Enclosed space at the after end of the ship, below the main deck. Used for ballast or fresh water.
Air holes.	— Small holes at the top of the floors and intercostals, required so that air may pass when the double bottom is being filled or emptied.
Air pipes.	— Goose necked ventilators to the double bottom.
Alleyway.	— A passage or corridor.
Anchor.	— Heavy iron implement used to hold the ship to a particular place in shallow water.
Anchor cable.	— Heavy chain used to attach the anchor to the ship.
Angle iron.	— Steel bar used as a stiffener in the ship's construction.
Awning spar.	— Wood spar placed overhead to support an awning.
Barrel.	— Main member of a capstan or winch to which a rope may be attached for the purpose of hauling on the rope. Cask or drum.
Beam.	— Transverse angle iron between opposite frames, provides strength and support to a deck.
Beam bolt.	— Bolt for locking a portable beam in position.
Beam knee.	— Bracket connecting a beam to a frame.
Beam socket.	— Socket for a portable beam.
Belaying pin.	— Loose pin in the sheerpole used to make fast running rigging on sailing ships.
Bilge.	— Rounded part at the bottom of the ship where the ship's side curves round towards the keel. This is the place in the hold where any loose water will collect.
Bitts.	— Strong twin posts for making fast mooring ropes.
Boat deck.	— Upper deck upon which the lifeboats are stowed.
Bollard.	— Strong single post for taking mooring lines.
Boom.	— A spar that can be rigged outboard for use with a patent log or a patent sounding machine, amidships, or for boats to tie alongside. A derrick.
Break.	— The place at which a fo'castle head or poop ends. Unfurl a flag after hoisting.
Breakwater.	— Vertical plate on foredeck, used to divert water coming aboard in heavy weather.
Bridge.	— High superstructure from which the ship is navigated.
Bridge deck.	— Deck above the main deck and beneath the bridge.
Bulk carrier.	— Vessel used to carry cargoes in bulk. (oil, grain).
Bulkhead.	— A vertical partition.
Bulkhead stiffeners.	— Angle irons used to strengthen a bulkhead.

Bulwark.	— Plating erected around the outboard edge of a deck.
Bunkers.	— Place in which the ship's engine fuel is stowed.
	Fuel for the ship's engines.
Butterfly nut.	— A nut with flanges attached for securing by hand.
Capstan.	— Vertical barrel used for hauling mooring ropes.
Cargo battens.	— Horizontal or vertical planks fixed to the inboard side of the frames, to protect cargo.
Ceiling.	— Fixed wood floor on the tank top underneath a hatchway, to protect the tank top.
Cells.	— Angle irons forming slots in the hold of a container ship, to hold the containers.
Chain locker.	— Compartment in the bow, above the fore peak, used to house the anchor cable.
Chains.	— A platform on either side of the ship's side, in the vicinity of the bridge. Used to swing the lead.
Chain plate.	— Deck plates to which the shrouds and swifters are secured.
Cleat.	— Metal lug on the side of a hatch coaming into which wood wedges are driven to secure tarpaulins.
	Piece of metal having two horns and used for securing a rope.
Coaster.	— Small ship engaged in the coastal trade.
Cofferdam.	— Vacant space between two watertight bulkheads, being the width of the ship, placed between engine room and oil tanks as a fire precaution or between oil and water tanks to prevent pollution.
Column.	— Samson post.
Companion way.	— Fixed staircase with handrails or bannisters.
Compressor.	— A mechanical device to hold the anchor cable or to hold a wire mooring rope and secure it.
	A machine for compressing air.
Counter stern.	— An overhanging stern.
Conventional ship.	— A ship equipped with masts, derricks and holds.
Cross-trees.	— A thwartship spar or table towards the top of a mast to which the derrick topping lift blocks are secured.
	Originally formed the support for the top mast.
Crow's nest.	— A look-out position towards the top of the fore mast.
Cruiser stern.	— A rounded stern without overhang.
Deadlight.	— Metal cover for a glass porthole.
Deck.	— Floor.
Deckhead.	— Ceiling.
Deck service line.	— Pipes carrying sea water for washing down, sanitary purposes and fire fighting.
Deep tank.	— A ballast tank the width of the ship with a centre fore and aft bulkhead, placed in either the 'tween-deck or the lower hold. Used for ballast or edible oils and in some cases provided with a large watertight lid enabling the space to be used for dry cargo.
Derrick.	— A pivoted boom, capable of being raised or lowered and swung from side to side. Used for loading and discharging cargo.

Devils claw.	– A two pronged hook. Used for securing the anchor cable while the vessel is at sea.
Docking bridge.	– Small thwartship bridge on the poop for use when docking.
Double bottom.	– Space between the bottom of the ship and the tank top. Used for the carriage of ballast, fresh water and oil bunkers. Subdivided fore and aft by the keelson and by a number of transverse bulkheads.
Draught marks.	– Marks on the stem and stern posts which will indicate the depth of water the vessel is drawing.
Drum end.	– Small drum on the side of a winch used for hauling on moorings and the discharge of cargo by fibre ropes.
Emergency aerial.	– Additional aerial for the use of the ship's wireless equipment.
Emergency lights.	– Spare oil navigation lights.–Battery lighting of the ship for use in an emergency.
Fair lead.	– Fixture which will lead a rope in a desired direction.
Fashion plate.	– Curved plate on ship's side at the break of the fo'castle head, or poop.
Feeder.	– Temporary wood structure fitted in the hatchway of a ship carrying grain in bulk.
Ferry.	– Small ship carrying passengers on a regular short voyage.
Fiddley.	– Space above the boilers on a steam ship.
Fish plate. (Curtain plate)	– Vertical iron plate on the outboard side of a deck in the superstructure. It adds strength.
Flair.	– Overhang of a ship's bows.
Flagstaff.	– Pole at the stern from which the ensign is flown.
Floor.	– Vertical iron plate running transversly along the bottom of the ship between opposite frames.
Flush deck.	– Said of a vessel whose weather deck runs from the bow to the stern, having no raised fo'castle head, bridge deck or poop.
Flying bridge.	– Bridge running fore and aft above the main deck on a ship to provide a safe passage for the crew in bad weather.
Fo'castle head.	– The uppermost deck in the bow of a ship.
Fore-foot.	– The rounded portion of the stem where it joins the keel.
Fore peak.	– A tank in the bow of the ship used for the carriage of ballast or fresh water, having a store room and the chain locker above it.
Frame.	– Vertical angle bar on the inside of the hull. The keelson, floors, frames and beams provide the skeleton on which the ship is built.
Freeboard.	– The distance that the statutory deck line is above water level.
Freeing port.	– Large opening in the bulwarks to allow excess water to escape quickly, in heavy weather.
Funnel.	– Exhaust pipe from the ship's engines. A chimney.
Gaff.	– Fore and aft spar on the after side of the aftermost mast. Inclined at about 45 degrees from vertical. Used to fly the ensign when at sea.

Term	Definition
Gallows.	– Iron framework on the fore and main topmasts, onto which the emergency oil steaming lights are hoisted.
	Iron framework for towing a trawl or paravane.
Ganger.	– Short length of chain cable between anchor and windlass.
Gantry.	– Main lifting frame of a crane.
	Three cornered frame fixed to the ship's side for the purpose of suspending the accommodation ladder.
Garboard strake.	– Plates forming the ship's bottom and which adjoin the keel or keel plate.
Gipsy.	– A sheave with interior lugs into which a chain will fit. Fitted to the windlass to take the anchor cable.
Goal posts.	– Two stump masts, one on each side of the ship, connected by framework at the top and with a topmast supported on the framework.
Gooseneck.	– Swivel at the heel of a derrick.
Gooseneck-ventilator.	– "U" shape at the top of a double bottom ventilator.
Hatch.	– An opening in the deck for the passage of cargo.
Hatch board.	– One of a number of wood or metal boards used to cover a hatch.
Hatch batten.	– Steel bar used to clamp a tarpaulin down at the sides.
Hatch coaming.	– Raised side of a hatch.
Hatch locking bar.	– Bar used to clamp the hatch-boards in position, fitted over the tarpaulins.
Hatchway.	– Hatch. Admits cargo to the hold.
Hawse pipe.	– Tube through which the anchor cable goes to the anchor.
Hold.	– A large cargo space the depth and width of the ship.
Houndsband.	– Band on a mast to which stays, shrouds and swifters are secured.
Hull.	– The whole of the ship's side plating together with the frames and floors.
Ice breaker.	– A specially built ship used for breaking ice.
Intercostal.	– A fore and aft plate, placed vertically between two floors.
Jackstaff.	– A flagpole in the bow.
Jolly boat.	– A working boat as apart from a lifeboat.
Jumbo derrick.	– A heavy lift derrick.
Keelson.	– A heavy angle iron running fore and aft the length of the ship, on top of the keel plate.
Keel plate.	– The centre fore and aft line of plates on the ship's bottom.
Kicking strap.	– A heavy prong pointing aft beneath a windlass gipsy, placed there to kick the cable out of the gipsy into the spurling pipe.
King beam.	– Portable beam in a hatchway with a flange in the centre of its upper edge, to enable it to take hatchboards on both the forward and after edges.
L.A.S.H.	– Lighter aboard ship system.
Lazerete.	– Storeroom.
L.G.C.	– Liquid gas carrier.
Lifeboat.	– Small open boat, provided for use in an emergency.
Lightening holes.	– Circular pieces are cut out of floors and intercostals, reducing the total weight of the ship.

Limber boards.	– Boards placed over the bilge space in the ship.
Load line.	– A vertical line on the ships side, beside the plimsoll mark, with horizontal lines running from it denoting the depth to which the ship may load under varying conditions and seasons.
Loop aerial.	– Aerial for the wireless direction finding equipment.
Lower bridge.	– Deck space beneath the navigating bridge.
MacGregor hatch.	– A patented steel hatch cover.
Magazine.	– A specially constructed place for the carriage of ammunition and explosives.
Main deck.	– The deck of the ship up to which all watertight bulkheads reach.
Main deck line.	– A line on the ships side above the plimsoll mark, denoting the position of the main deck. It is from this line that freeboard is measured.
Man-hole.	– A circular hole cut in a tank, tank top, or other place, through which a man may climb.
Man-hole cover.	– Must be bolted down and watertight.
Margin plate.	– Plate running the length of the bilge and forming the outboard boundry of the double bottom tanks.
Mast.	– Vertical pole in the centre line of the ship.
Mast house.	– A house at the base of the mast, on deck.
Mast table.	– The top of a mast house or a platform in the same position.
Monkey island.	– Top of the wheelhouse.
Monkey's face.	– Connecting piece of a mooring swivel.
Motor ship.	– Ship driven by internal combustion engines.
O.B.O.	– Oil-Bulk-Ore carrier.
Orlop deck.	– The lowest deck.
Panama lead.	– Fairlead with a closed top. Used in the Panama Canal.
Pillars.	– Strong vertical angle irons supportintg the corners of the hatchways, above the tank top.
Plimsoll mark.	– A circle with a horizontal line running through the centre, on the ships side. Marks the maximum depth to which the ship may load in salt water in summer.
Poop.	– Weather deck at the stern of the ship.
Port.	– Circular window. Harbour.
Portable beam.	– Loose transverse strengthener in a hatchway that may be removed when working cargo.
Portable rail.	– Loose rail on the ship's side.
Propeller.	– Three or four blades, each being part of a screw thread. Propels the ship through the water.
Quadrant.	– Large quarter circle on the rudder post head.
Radar.	– Equipment that shows on a screen any obstacles in any direction.
Radar scanner.	– A horizontal revolving aerial for the radar.
Rat guard.	– A circular piece of metal that fits on mooring ropes, to prevent rats boarding or leaving the ship.
Relieving tackle.	– A tackle fitted to the quadrant for the purpose of absorbing rudder shocks in bad weather.

Deck line.

Load lines:
TF, F, T, S, W, WNA

Forward →

Plimsoll mark (R, L)

Timber load lines:
LTF, LF, LT, LS, LW, LWNA

Tonnage mark.

Layout of a ship's marks.

Term	Definition
Rigging.	— Any ropes attached to a mast or on the gear attached to any mast.
Rolling chocks.	— Angle bars placed on the outboard side of any ship, in the vicinity of the bilges, running fore and aft at right angles to the hull.
Roll on/Roll off (Ro-Ro).	— A vessel equipped to allow vehicles to drive aboard, drive ashore.
Rolling beam.	— Portable beam that is moved forward or aft on rollers.
Rudder.	— Means by which the direction of the ship's head is controlled. A moveable flat plate at the stern.
Samson post.	— A stump mast used to support a derrick, also known as a King post or Column.
Sanitary tank.	— A tank filled by the deck service line which supplies sanitary water by gravity.
Satellite navigation-antenna.	— Aerial for the satellite navigation equipment.
Screw.	— Propeller.
Scupper.	— A drain.
Seaway lead.	— Special fairlead for use in the locks on the Great Lakes. Sometimes called "Port Colborne lead".
Sheerpole.	— Horizontal pole passed through the thimbles of the shrouds and swifters at their lower end, to prevent the wires turning and unlaying. No longer in common use.
Sheer strake.	— Line of plates on the hull, whose top edge is attached to the main deck.
Shelter deck.	— Covered deck space above the main deck.
Shifting boards.	— Wood or metal boards temporarily erected vertically, fore and aft in the centre of a hold to prevent grain or other unstable cargo from moving.
Side door.	— A watertight door in the ship's hull through which passengers, stores or cargo can be embarked.
Side lights.	— The red and green, port and starboard navigation lights.
Signal mast.	— A mast to which various coloured lights are attached which can be switched on in various patterns as signals. Otherwise known as "The Christmas Tree".
Skylight.	— A glazed opening in the deck that allows light to pass to the deck below. Often raised and hinged so that it may be opened in fine weather for ventilation.
Sounding pipe.	— A pipe leading from the weather deck to a bilge or double bottom tank, down which a sounding rod may be passed to ascertain the amount of water in the compartment.
Spar ceiling.	— Cargo battens.
Spider Band.	— Head band on a derrick, made in two parts. (Originally fitted at the base of the mast to carry belaying pins).
Spurling pipe.	- Pipe by which anchor cable enters the chain locker.
Stabilisers.	— Horizontal plates that can be pushed out of the hull in bad weather to increase the waterplane and reduce the roll of the ship. Withdrawn to enter port.
Stanchion.	— Vertical post which helps to support the deck above.
Steaming lights.	— The two white mast head lights used for navigation.

Term	Definition
Steering flat.	— The position in the ship in which the steering engine is located.
Stem.	— A vertical angle iron rising from the keel and to which the bow plates of the hull are fastened.
Stern light.	— A white light at the stern used for navigation.
Stern post.	— A vertical angle iron at the stern of the vessel, carries the weight of the rudder.
Stokehold.	— The position in the ship from which the boilers on a steam ship are fired.
Striker plate.	— A doubling plate on the ship's bottom, beneath a sounding pipe, to take the strike of the sounding rod.
Stringer.	— Longitudinal, horizontal plates along the ship's side, inboard, that in conjunction with the frames give strength to the vessel.
Strong room.	— A specially constructed compartment for the carriage of mails and bullion.
Strum box.	— A perforated metal box on the end of a suction pipe, placed there to prevent dirt entering the pump. Otherwise known as a rose box.
Stulken derrick.	— A patented heavy lift derrick of German origin.
Suction pipe.	— A pipe used for the drawing of fluid from a place to a pump.
Summer tanks.	— Small tanks on the outboard sides of a tanker.
Taffrail.	— Upper rail going around the stern.
Tanker.	— Vessel constructed specifically to carry liquid cargo.
Tank top.	— The bottom of a hold which is also the top of the double bottom tanks.
Tarpaulin.	— Large sheet of waterproof canvas used to cover a hatch and make it watertight.
Telescopic topmast.	— A topmast that will lower into the mast.
Telegraph.	— Equipment (other than a 'phone) for conveying orders from the bridge to engine room or poop.
Timber load line.	— Additional load line for use with timber cargoes.
Tonnage hatch.	— A small hatch in the weather deck of a ship which cannot be properly battened down, when the main deck is below the weather deck. Originally used to avoid tonnage dues. Now becoming obsolete.
Tonnage mark.	— Triangular mark on the side of the ship adjacent to the plimsoll mark.
Top mast.	— A short light mast placed on top of the mast.
Trimming hatch.	— Small hatch placed in the corners of 'tweendecks, for the use of stevedores when trimming bulk cargo.
Truck.	— Flat disc on the top of a topmast.
Trunnion.	— Swivel fitting on the heel of a jumbo derrick.
Tug.	— Small vessel used to manoeuvre large ships when entering and leaving port.
Turbine ship.	— Steam ship driven by turbine engines.
Tumble home.	— The inward inclination of a ship's side between her maximum beam and the deck.
Tumbler.	— Fixture on the cross trees to which a topping lift block is secured.

Tunnel.	– Steel tunnel, housing the propeller shafting between the engine room and stern gland.
'Tween-deck.	– Any deck below the main deck.
Union plate.	– Triangular plate with a hole in each corner.
VLCC	– Very large crude oil carrier.
Ventilator.	– Any arrangement that allows the air in a compartment to be changed.
Ventilator plug.	– Wood plug or metal lid used to close a ventilator shaft and render it watertight
Washport.	– Freeing port.
Waterplane.	– The area of a ship's side that is under water.
Watertight door.	– A door lined with packing and screwed down between two watertight compartments. Must be closed when the ship is at sea. (This does not apply to the tunnel door which can be closed from the upper deck).
Wedge.	– Triangular shaped piece of wood or metal used to keep hatches closed and watertight.
Well.	– Depression in the tank top which is used to collect any water in a hold, when the ship has her double bottom extended to the ship's side. (Sometimes called a Hat box)
Well deck.	– That part of the weather deck between the break of a raised fo'castle head and the bridge deck, or the bridge deck and the break of a raised poop.
Wheelhouse.	– Superstructure on the navigating bridge which houses the navigating and steering equipment.
Winch.	– Machine having a horizontal barrel operated by either hand or power, to which a rope may be made fast and wound around the barrel. The machine will, by rotating the barrel, cause the rope to haul or hoist an object.
Wind chute.	– A metal scoop that fits into a port and projects outward, in order to catch wind and assist ventilation.
Windlass.	– Machine designed primarily to raise the anchor from the sea bed.
Yard.	– Transverse horizontal spar, situated near the top of the top mast. Originally one of a series of spars on which square sails were set.

CONSTRUCTION.

A merchant ship is constructed for the sole purpose of transporting goods from one place to another by sea. It has an engine room to house an engine that turns a screw or propeller which thrusts the hull forward through the water and a moveable plate (rudder) at the rear, for directing its path.

Basically to build any ship, a line of horizontal plates called the keel plate forms the base, on top of the keel plate is a very strong vertical steel plate, the keelson. This forms the backbone of the ship and curves upwards at the front to form the stem and upwards at the back to form the stern post. Stout vertical girders of the same height as the keelson are set at right angles to the fore and aft line of the ship, these are the floors. Spaced at intervals between the floors and parallel to the keelson, are more stout vertical girders, these are the intercostals. The point at which the hull plating will curve from horizontal to vertical is the bilge.

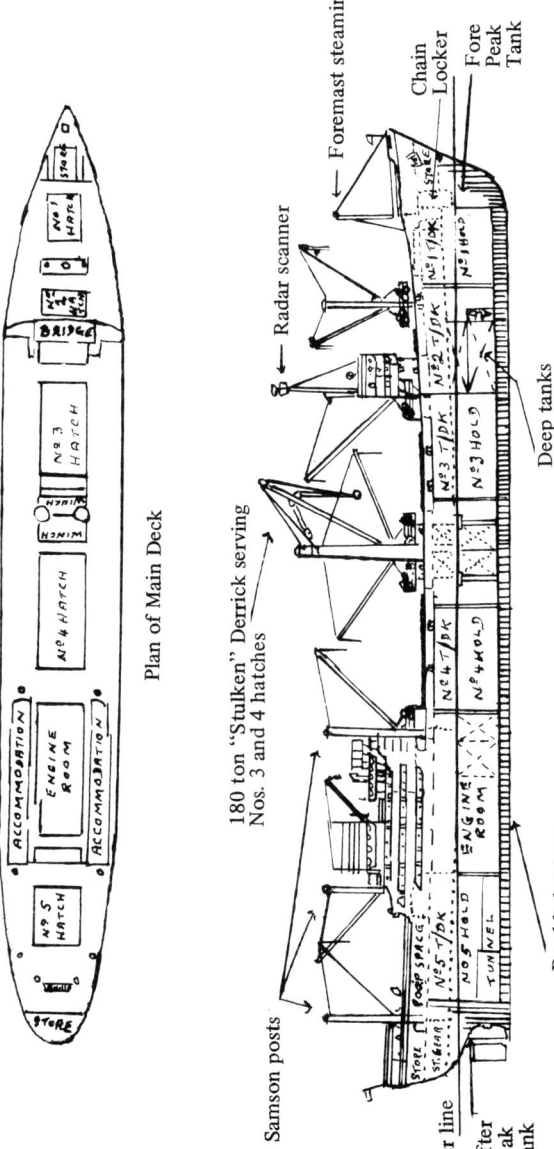

Plan of Main Deck

180 ton "Stulken" Derrick serving Nos. 3 and 4 hatches

Plan of the Heavy Lift Vessel "Adventurer" Messrs. T. and J. Harrison Ltd.

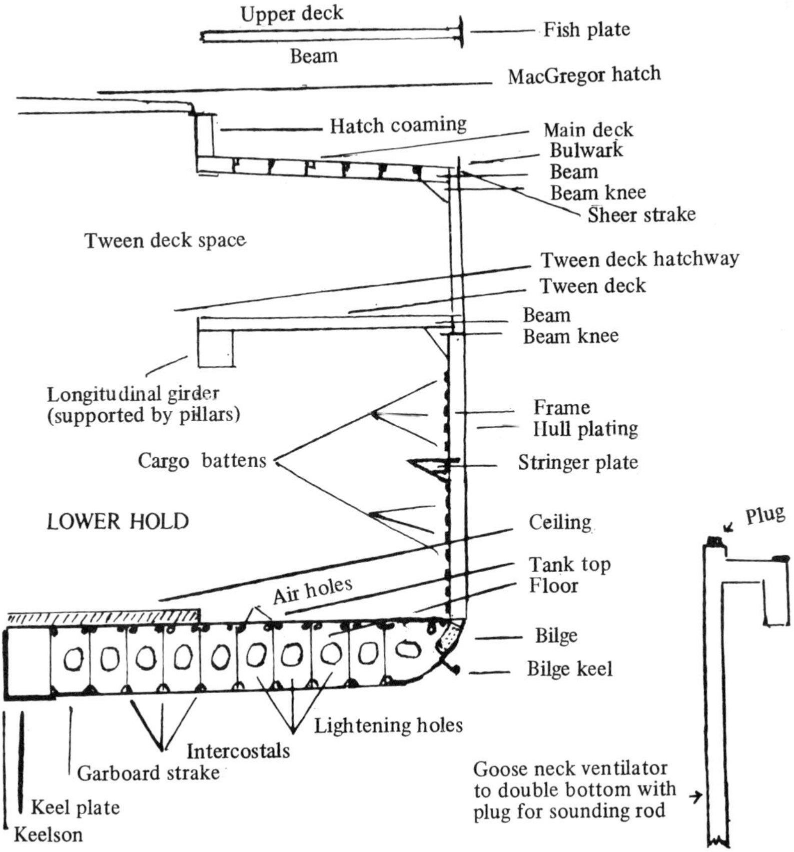

Cross Section of a Typical Conventional Ship.

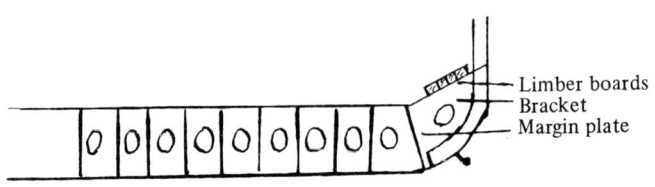

An Alternative Type of Double Bottom

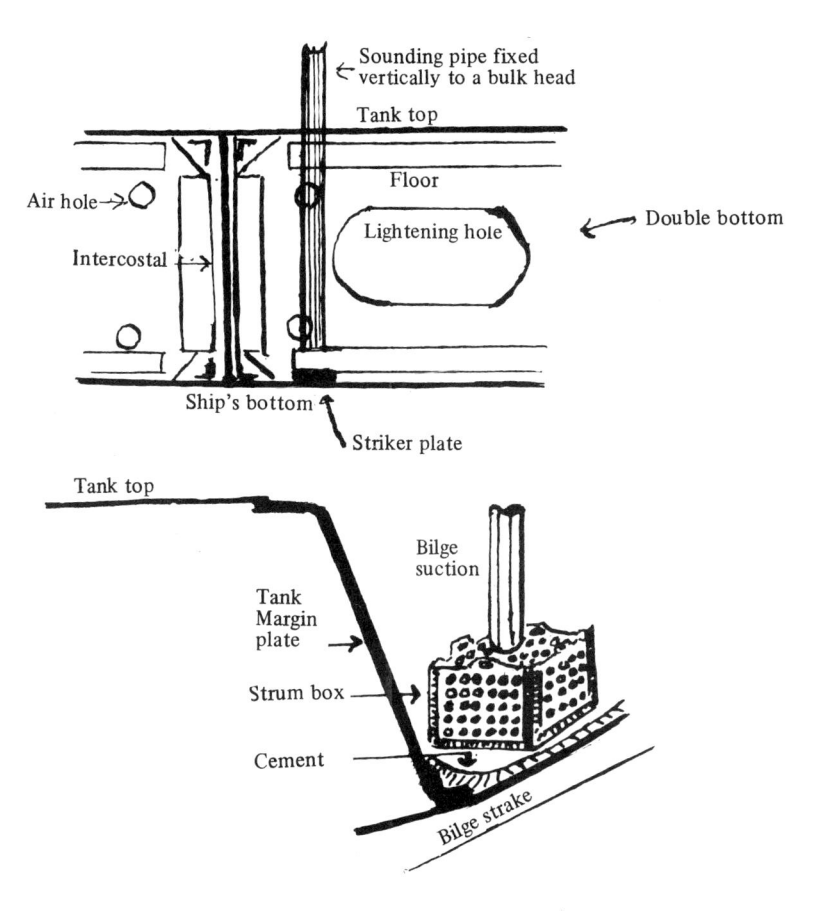

Typical strum box in a bilge. The sides being loose so that they can be removed for cleaning.

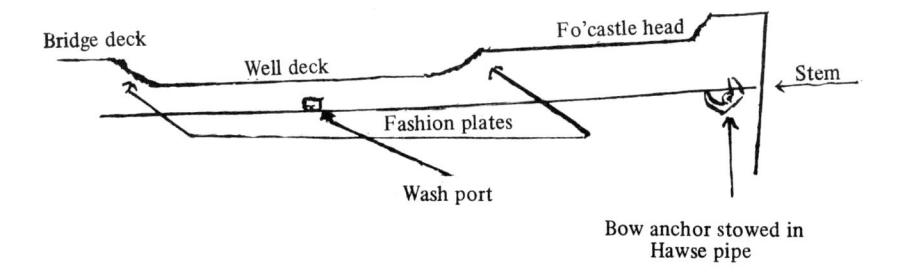

Here the floors end and are joined to vertical angle irons called frames by brackets. These frames form the skeleton onto which the hull plates are either riveted or welded.

On all ships except oil tankers, the floors and intercostals are covered with steel plating which forms the bottom of the hold or cargo space, this plating is called the tank top. The ship now has a cellular double bottom, which is divided into double bottom tanks. These tanks may stretch to the hull plating either side of the ship but very often the tank ends at the bilge, a little way short of the hull plating, in these cases, the vertical fore and aft plate confining the limit of the double bottom tank is called the margin plate. The space between the margin plate and the hull plating is the bilge and it is used to collect any water that may enter the hold. The bilge space is covered with wood boards called limber boards which are portable and can be moved to give access to the bilge for cleaning. Outside the hull, in the way of the bilge, fitted at right angles to the hull and parallel to the keelson is an angle bar, called the bilge keel, or rolling chock.

Inside the hull plating at vertical intervals, horizontal plates the length of the ship are provided for strengthening purposes, these are stringers.

At their top, the frames on each side will be joined together, across the width of the ship by angle irons called beams. The beams are covered by plating to form a deck. The underside of any deck is known as a deck-head. The deck at the top of the frames is usually, though not always, the main deck. It is also of course the weather deck.

At various intervals, along the length of the ship, vertical steel walls running from the main deck to the ship's bottom plating and from the frames on one side of the ship, to the frames on the other side are provided to break the ship up into several watertight compartments. These are the watertight bulkheads and the angle irons running vertically at intervals to give additional strength are the bulkhead stiffeners.

Above the weather deck are various houses forming among other things, the bridge and the accommodation. The whole of this is termed the superstructure.

The fore and aft line of hull plates next and parallel to the keel plate, on each side of the keel plate, is the garboard strake. The fore and aft line of plates at main deck level, is the sheer strake. Shaped hull plates that have been cut at one end with a curve, where the hull plates rise to another deck, are fashion plates. Narrow plates that run along the outboard boundry of any deck above the weather deck are fish plates and are placed there to give added strength.

Access to the double bottom tanks is provided by man-holes, cut in the tank top, man-hole covers are both water tight and bolted down. The floors and intercostals are lightened by having oval pieces cut out and the holes are large enough for a man to crawl through for the purpose of inspection. The plates are also cut to allow the passage of air at the top and fluid at the bottom so that the tanks can be filled and emptied.

From this point on, the structure of the ship will be determined by the cargoes (goods) she is to carry and the ports to which she is expected to trade.

Conventional Cargo Vessels have large openings in the upper deck and in any decks below the upper deck. These openings (hatchways) on the upper deck, will be bounded by vertical plates not less than 2 feet 6 inches (76cm) high, which are the hatch coamings. Any deck below the main deck, is a'tween-deck, when there is more than one 'tween-deck, the lowest deck above the tank top is termed the orlop deck. Underneath the hatchway, the tank top is often protected by a wood platform called the ceiling. Hatchways in 'tween-decks will have little or no coaming but are provided with portable stanchions and guard chains in lieu.

The frames will all be lined with a series of either horizontal or vertical boards known as cargo battens or the spar ceiling. Cargo battens are provided to keep dry cargo off the frames and hull plating, thus preventing damage to the cargo by condensation.

The ends of the beams that are cut to allow for the hatchway opening, will be supported by a strong girder, which in its own turn is supported by the floors underneath the tank top, by means of either two or four pillars. Before the hatchways are closed, portable beams are shipped between the ends of the cut beams, to help maintain transverse strength, in the break of the hatchway.

Many conventional ships with 'tween-decks have a small hatchway cut in each corner of the 'tween-decks. These trimming hatches are provided for the use of dockworkers and others who may have to trim a bulk cargo such as grain or coal, up underneath the deck-head.

Again, many conventional ships have large tanks placed in one or more of the lower holds or 'tween-decks, either completely across the ship, or in the wings. These deep tanks are provided for the carriage of edible oils and may also be used for sea water ballast. Very often they have large watertight lids which can be removed so that tanks can also be used for the carriage of dry cargo.

These ships will also be supplied with derricks or possibly cranes for the purpose of handling cargo, in ports where this is necessary. Heavy lift ships are provided with at least one very strong derrick capable of lifting anything from 30 to 300 tons according to the needs of the trade in which the ship is employed.

Passenger ships are much the same as conventional ships but they have more watertight bulkheads and therefore more watertight compartments and more decks, all this, apart from being useful for the carriage of passengers, gives them a greater safety factor.

Bulk carriers will be built to suit the carriage of a particular type or types of bulk cargo. The holds may be shaped to avoid the need for trimming, or to ensure that a portion of the cargo can be loaded at a higher level (this applies mainly to ore carriers and helps stability). They may carry special equipment for the rapid discharge of their cargo by means of the ship's equipment.

The term bulk carrier includes not only dry cargo carriers but also gas and oil tankers.

Gas tankers are constructed in very much the same way as the conventional ship. Each hold contains a cargo tank, which is in no way secured to the ship's structure but is held in place with special chocks. The space between the hull, deck-head and the tank top at the bottom of the hold is usually filled with an inert gas. There are of course no hatchways or derricks (except those used to lift the cargo hoses and stores aboard). Cargo is pumped aboard or ashore by means of hoses and pipelines.

Oil/Bulk/Ore carriers have double bottom tanks which are used solely for the carriage of fresh water, salt water ballast and bunkers. They will also be provided with wing tanks which may be used to carry either oil or a bulk cargo such as grain. The holds which have oil tight hatch lids may be used for the carriage of

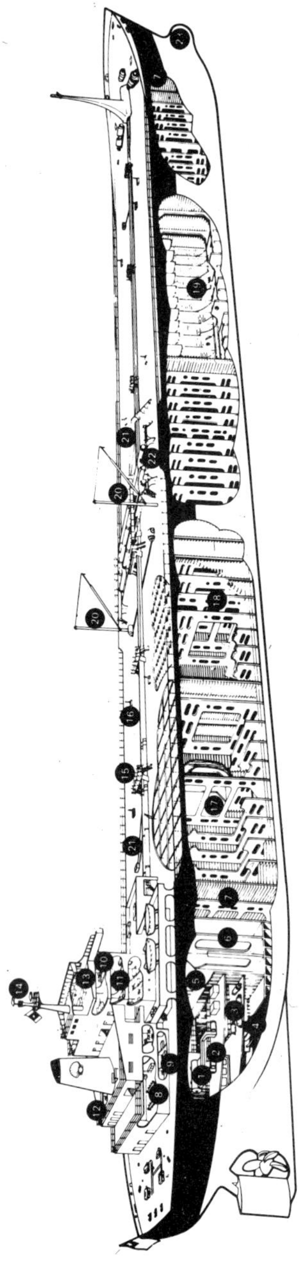

A modern oil tanker

Length overall	1076 ft
Breadth (extreme)	155 ft
Depth (extreme)	80 ft 6 in
Dwt	208 000 (approximately)
Draught loaded	62 ft 3 in
Engines	Single screw, geared turbine designed for 27 500 shp
Speed	$15\frac{1}{2}$ knots (service speed)

By courtesy of Shell Tankers (UK) Ltd

1 Boiler
2 Force-draught fan
3 Auxiliary machinery
4 Water inlet pipe
5 Control room
6 Oil fuel bunker
7 Permanent ballast tank
8 Crew mess
9 Crew accommodation
10 Master's accommodation
11 Officers' accommodation
12 Swimming pool
13 Navigating bridge
14 Radar scanner
15 Fire hydrant
16 Cargo hatch and stand pipe
17 Structural arrangement – centre cargo tank
18 Structural arrangement – wing cargo tank
19 Structural arrangement – wing cargo tank
20 Hose and cargo derrick
21 Breakwater
22 Loading/discharge pipelines
23 Bulbous bow

TYPICAL SECTION I.W.O. L.P.G. TANK.

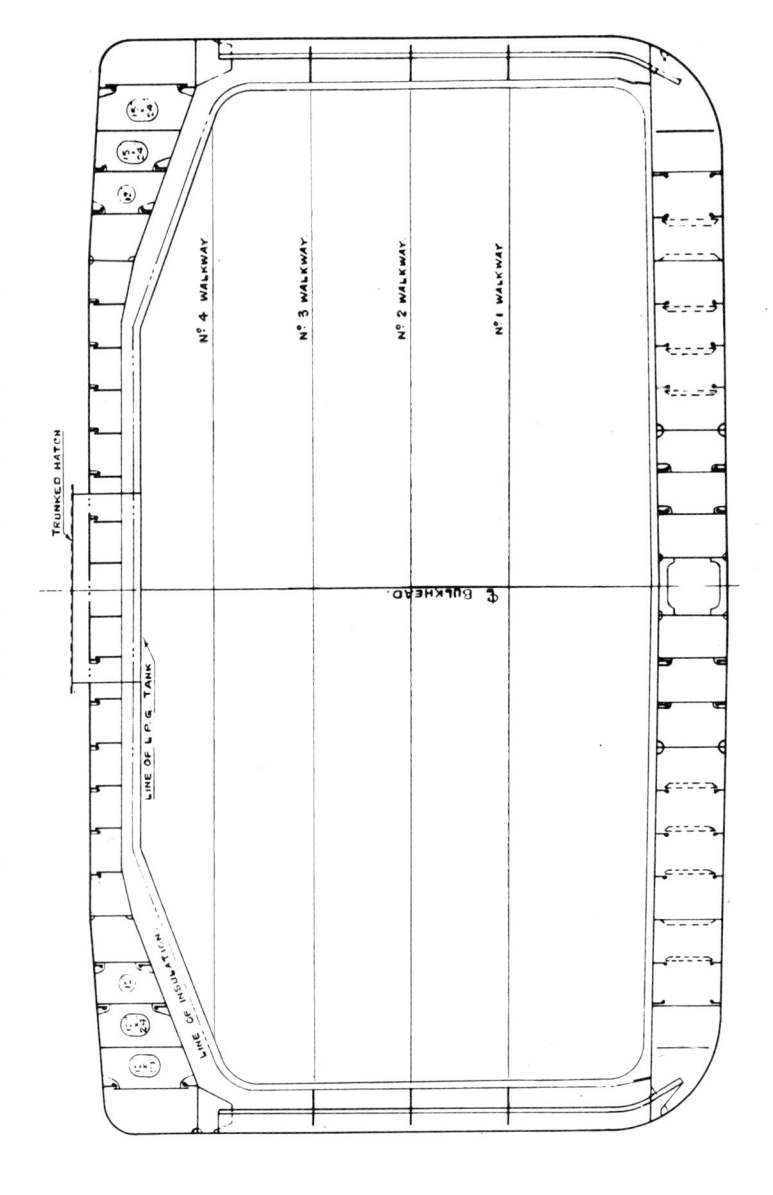

M.S."AMRA" 10,024 G.T. at the Royal Albert Dock London, loading six 45 ton lighters (75' x 18') for the port of Mogadiscio in Somali with her 300 ton Stulcken heavy lift derrick.

"Encounter Bay" loaded in full swell off the Hook of Holland. Each ship is capable of carrying over 1,500 standard 20' containers, 304 of which are insulated.

"Roll on/Roll off"

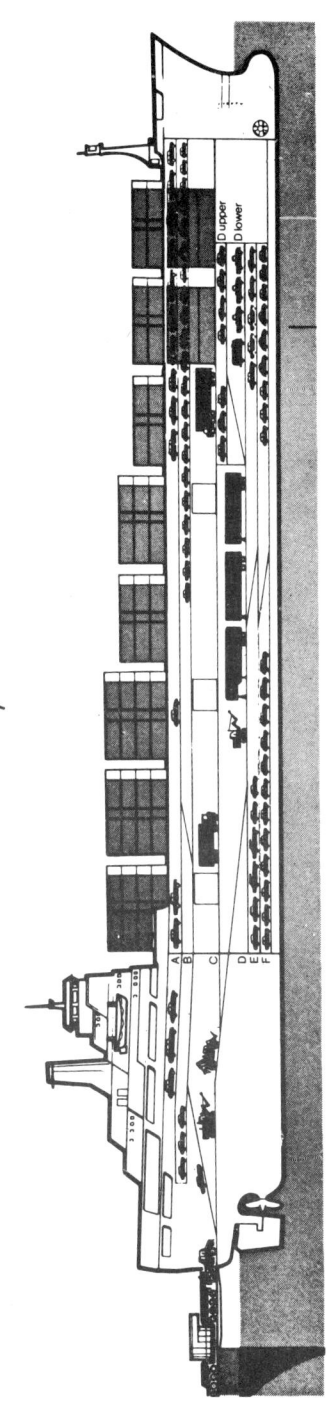

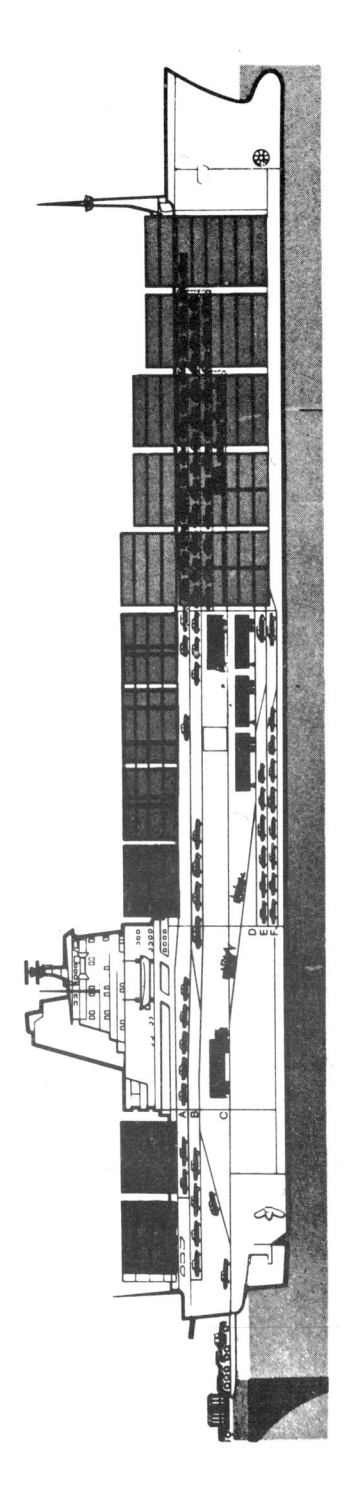

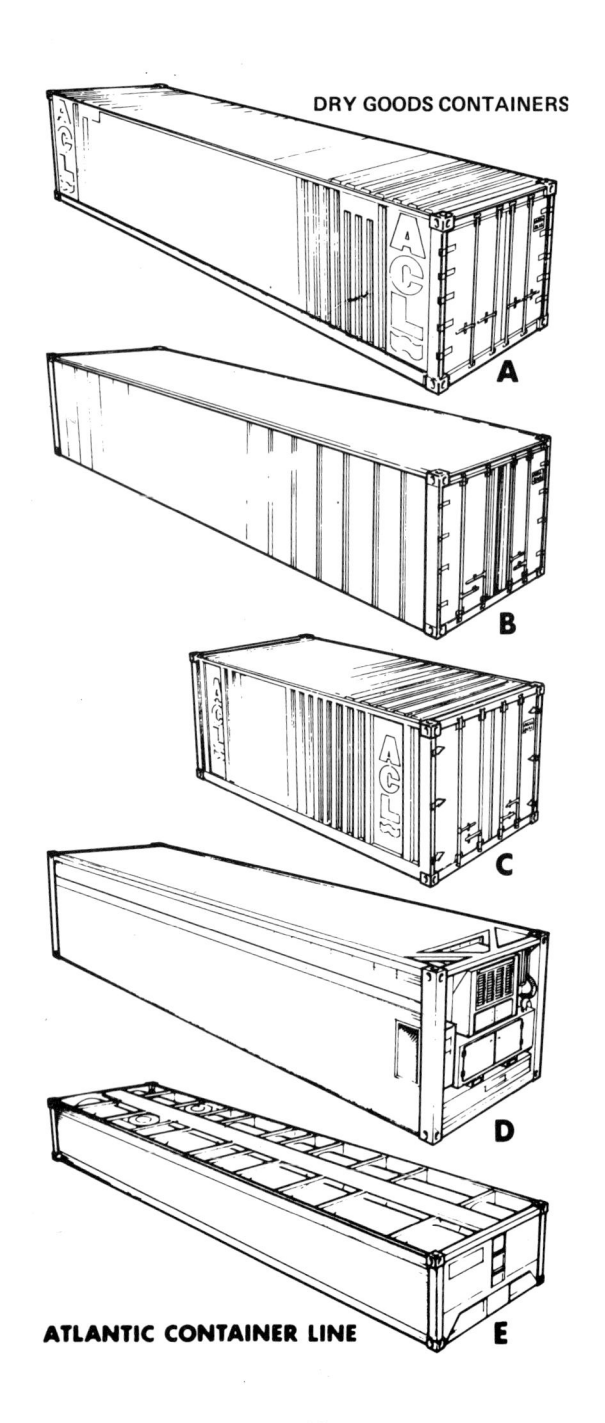

DRY GOODS CONTAINERS

A

B

C

D

ATLANTIC CONTAINER LINE E

A.C.L. ROLL–TRAILERS

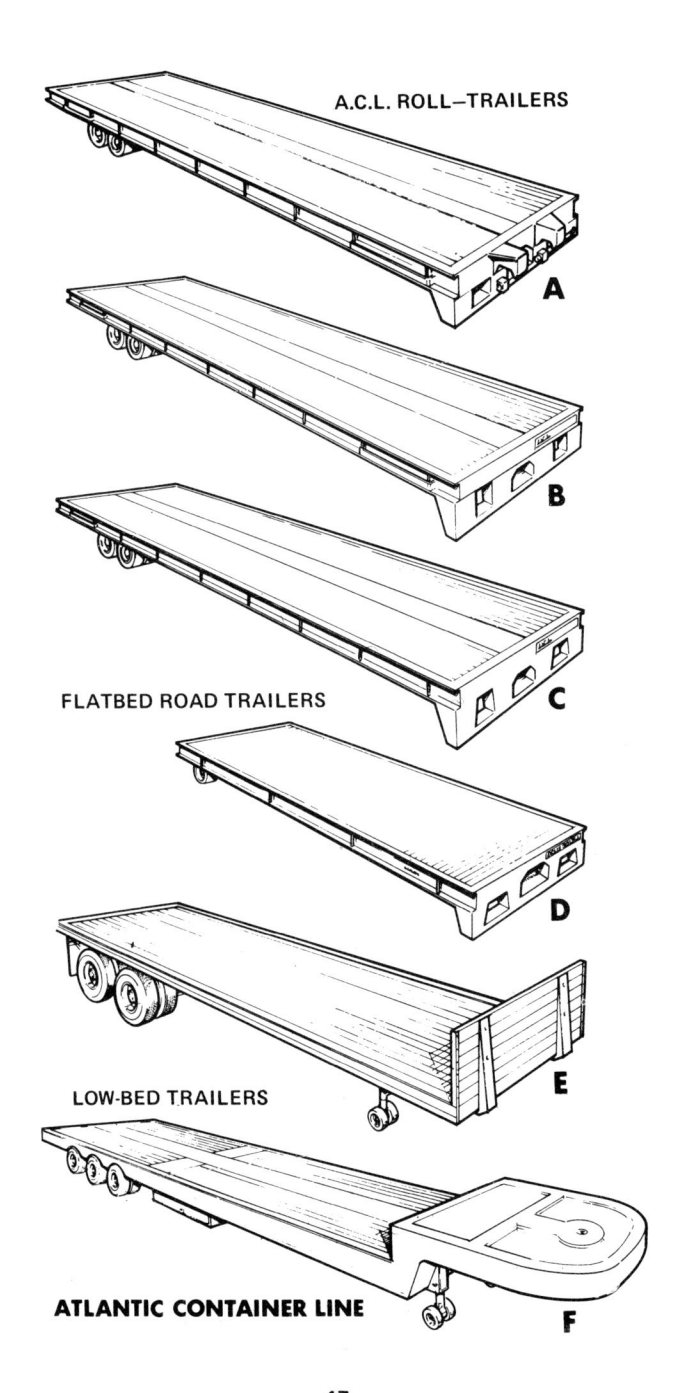

A

B

C

FLATBED ROAD TRAILERS

D

LOW-BED TRAILERS

E

ATLANTIC CONTAINER LINE

F

47

m.v. 'Anglo Norness'

924 feet
231 feet
1,099 feet
176 feet
90 feet

oil, bulk grain or ore. They can carry a full cargo of oil or a full cargo of ore or bulk grain or a cargo which is a mixture of these. The 258,000 ton dwt "Anglo Norness' oil/ore carrier pictured here is one of the biggest of her kind at the time of going to press and as shown four "Jumbo Jets" could be carried on her decks. The cross section shows how the cargo is stowed.

Oil tankers have no double bottom tanks but are divided up into numerous self-contained tanks, each tank has its own hatch and tank lid on the main deck. Cargo is transferred by means of pumps, hoses and pipe lines.

LASH Lighter aboard ship system. In this system, ships fitted with heavy lifting gear carry a cargo of fully loaded lighters to a main port. The discharged cargo of lighters is then sailed under its own power to the various out ports. Meanwhile lighters which have loaded in the out ports are collected at the main port for shipment elsewhere.

Container ships, constructed as such, have vertical steel girders fixed in the holds to act as cells. The containers which are of a standard size are slotted into these cells, when the holds are full additional containers are carried on deck. Wing tanks occupy the space between the cells and the ships side and these may be ballasted as required to keep the ship on an even keel.

Roll on/Roll off (Ro-Ro) ships are designed to allow cargo in the shape of vehicles to be driven aboard over a ramp lowered from either the stem or stern of the ship or from a side door or doors and from upper to lower decks (and vice-versa) by means of internal portable ramps. The internal ramps which also act as hatch covers in the various 'tween-decks will, when closed, carry cargo in the same way as cargo is carried on conventional hatch covers.

These ships may be part Ro-Ro and part containerised or conventional and are built according to the expected requirements of the trade in which they are intended to operate.

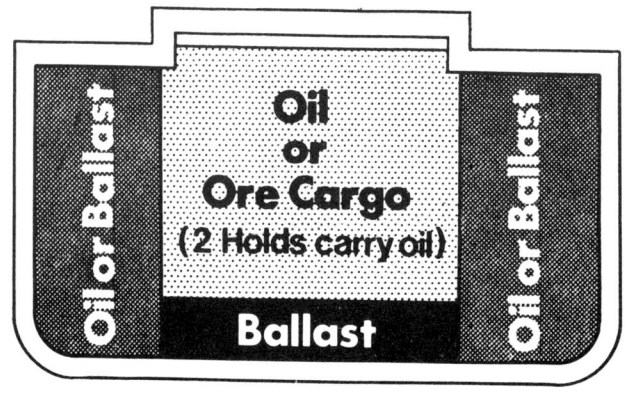

Cross section "Anglo Norness"

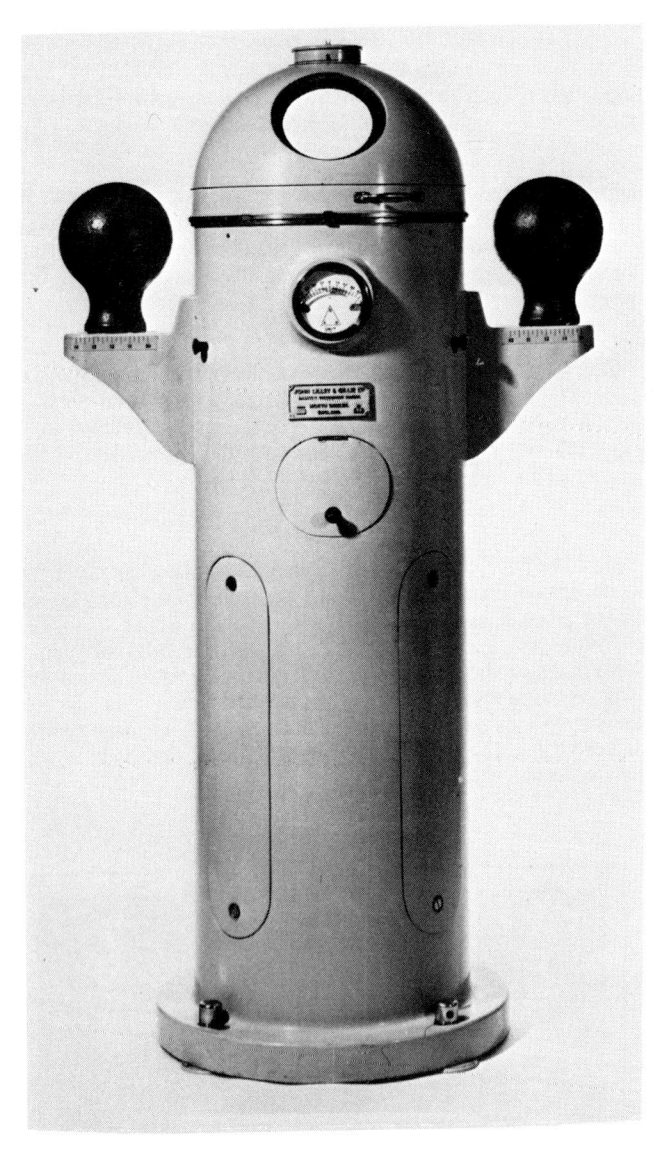

A typical magnetic standard compass binnacle.

THE MARINERS COMPASS

CHAPTER 4.

COMPASSES, HELM ORDERS AND STEERING GEAR.

Compasses.
A compass is an instrument that determines in which direction either the true or magnetic north pole lies. There are in fact two completely different instruments, both of which are compasses and are capable of accomplishing this feat.
These compasses will indicate (after correction) the direction of true north and as a result, all other directions, by means of a graduated circular compass card similar to the one illustrated here.

The magnetic compass consists of a graduated circular card marked both in degrees from 001° to 360° and with the 32 points of the compass (11¼ degrees being equal to one point) and sometimes to quarter points. This card will have an even number (2, 4 or 6) of long thin magnetic needles attached to it, and will be suspended from a cap in the centre, bearing on a pivot. The whole being fitted into a glass topped bowl. The bowl is then suspended horizontally from two points by triangular supports called gymbal pins which bear on a ring called a gymbal. The gymbal in its turn is suspended at two points lying at right angles to the gymbal pins in a stand known as a binnacle. No matter what movement is made by the ship in a seaway, the compass card and the bowl that contains it, will by swinging in the gymbal, remain level.
One end of the magnetic needles is attracted to the magnetic north pole and as they are free to rotate, carrying the card with them, round the central pivot. The direction in which the magnetic north pole lies becomes immediately apparent.
Unfortunately, this is not sufficient for the purpose of navigation, because the position of the magnetic north pole is not exactly the same as the position of the true north pole. The angle between the differences in the directions of the magnetic and true north poles, although it varies considerably from place to place, is known and the variation as this error is called, can be allowed for when calculating the direction in which the true north pole lies.
Ships are made of iron and steel and the metal in the ship also has an attraction for the magnetic needles in the compass. In order to reduce this attraction as much as possible, compensating magnets are placed in the binnacle. Naturally, they are not completely effective but when these magnets are properly adjusted, they do keep the deviation, as this error is called, within bounds.
Moreover, it is always possible with the aid of tables, to find the exact error of the compass, by taking a bearing of the sun or a star, when it is fairly low on the horizon, or without tables, by taking a bearing of the North Star.
Every ship is required to have a magnetic compass. This compass, known as the Standard Compass is regularly checked by the Officer of the Watch for its error and then compared with other compasses on the ship, to ensure that they are in fact correct.
A magnetic compass that is specifically used by the helmsman to steer a course, will have the bowl filled with liquid. This is done to dampen the motion of the card and help prevent the card swinging either with the motion of the ship

or from some other cause. This type of compass is known as a liquid compass.

When the helmsman is supplied with a gyro repeater, for the purpose of steering the ship, a separate magnetic compass is not necessarily supplied but if it is not, there will be an arrangement by which the helmsman is able to use the standard compass for this purpose, if it becomes necessary.

There will also be a binnacle or pedestal placed aft, into which a compass may be shipped for use with the emergency steering gear.

Both the standard and steering magnetic compass binnacles are placed on the centre line of the ship. Inside the compass bowl, where it can be seen and in line with the fore and aft line of the ship, is a black line. This line known as the lubber line, will indicate on the compass card the direction in which, according to the compass, the ship is heading. When this reading is compensated by the known error of the compass, the true direction in which the ship is heading is determined.

Owing to the attraction of iron or steel toward the magnetic compass needles, no iron or steel should be allowed to come in close proximity to the compass. Helmsmen should not carry knives or spikes on their person when steering. Transistor radios, too, have an alarming effect on magnetic compasses.

The Gyro Compass is an electrical machine. It works on the principal of a childs top rotating at a pre-determined constant speed, with the result that the axis of the top remains constantly pointing in the same direction as the earth's axis and therefore towards the true north pole, with possibly some slight known error.

With one master compass, as many repeater compass cards as may be required, can be placed in different positions about the ship. One is supplied for the use of the helmsman, the card may be upright and much of it may be obscured but it will in fact be graduated in the same manner as a magnetic compass card. The helmsman being required to steer by degrees. A pointer will take the place of the lubber line and indicate the course upon which the ship is heading.

Where an automatic helmsman is installed, it will usually operate from a gyro compass, however, an automatic helmsman can be installed to operate from a magnetic compass, if this is required by the shipowner.

Should a gyro compass repeater appear to act at all strangely or should the difference between the gyro course and the magnetic course on the steering compass alter, the gyro is immediately suspect and the attention of the Officer of the Watch should be called to it, so that he may determine if in fact the gyro compass or the repeater, is in good working order.

A good helmsman supplied with a gyro repeater, will always check his course with the magnetic compass occasionally, in order to ensure that the gyro is functioning correctly.

It is not intended that the reader should come to the conclusion that a gyro compass is an unreliable instrument. Far from it but any instrument is liable to failure from one cause or another and usually this occurs at an unexpected moment. The only remedy, apart from meticulous servicing, is to keep a constant check on any scientific instrument when it is in use.

The Arma Brown Gyro Compass.

Donkin duplex ram hydraulic Telemotor with automatic by-pass.

Donkin Console incorporating a Donkin hydraulic Telemotor and autopilot controls.

Donkin pump type hydraulic Telemotor.

Automatic and hand steering control with gyro compass repeater by Messers. S.G. Brown.

Wheelhouse M.V. 'Wiltshire', Messrs. Bibby Lines Limited

Steering gear.

The requirement of any steering gear is that it shall move the rudder to the exact position required by the helmsman and keep it there until the helmsman requires a further movement of the rudder. The gear must be able to act both quickly and easily and without causing the helmsman undue fatigue.

Small ships and tugs may be supplied with hand steering gear, in which the power required to move the rudder is supplied, perhaps through gears, by the helmsman. It is more usual however for small craft to have powered steering gear which can be coupled to, or uncoupled from, the hand steering gear as required. The powered gear being supplied for use in narrow waters, while the hand steering gear is used in the open sea.

Some small craft together with some of the older ships, may be fitted with rod and chain steering gear. The wheel on the bridge is connected by a series of rods to a steering engine located at deck level, in the after end of the engine room. The engine drives either one or two drums or gipsy wheels which wind chain on one side and off the other. The chains, coupled to rods, are led along the deck to the tiller or quadrant. Buffer springs are incorporated to absorb shock.

Up to date methods use either electric impulses or an oil filled hydraulic system known as tele-motor to connect the wheel with the engine. The engine is situated at the rudder head and may be either steam or electric but in either case the rudder head will be moved by hydraulic power. One type of engine has either two or four cylinders containing rods that are connected to the rudder head. When the helmsman turns the wheel, he operates what is known as the hunting gear. The hunting gear causes hydraulic pressure to act on the rams, as the rods are called, so that the rudder is forced round to the required position. This is called hydraulic ram steering gear. Another type is the rotary vane steering gear, in this, vanes are attached to the rudder head and the whole is enclosed in a chamber filled with oil. When the helmsman turns the wheel and operates the hunting gear, oil is forced into one side of the chamber and brings pressure to bear on the vanes attached to the rudder head, forcing them round until the rudder is in the required position. In an emergency, when for some reason the steering gear has broken down, it is possible in both these systems to operate the hydraulic system by means of a hand pump, thus obviating the necessity of rigging tackles from the quadrant to winch drums for the purpose of controlling the rudder.

When the tele-motor equipment is used to connect the wheel to the hunting gear on a steering engine, the wheel will tend to return automatically to midships when released, this in turn allows the hunting gear to return the engine and rudder to the midship position. Latterly, some ships fitted with tele-motor equipment, have been fitted with a rotary pump in the tele-motor system. This removes the tendency of a released wheel to return to the midship position and the wheel remains standing in any position in which it is left. However, under these circumstances, the relationship between the positions of the wheel and the rudder is not always exactly maintained, so that it then becomes necessary to fit an electric rudder indicator, to indicate the exact position of the rudder at any given moment.

Relieving tackle suitable for a small vessel.

Secure a handy billy to each hauling part, heave tight and make fast.

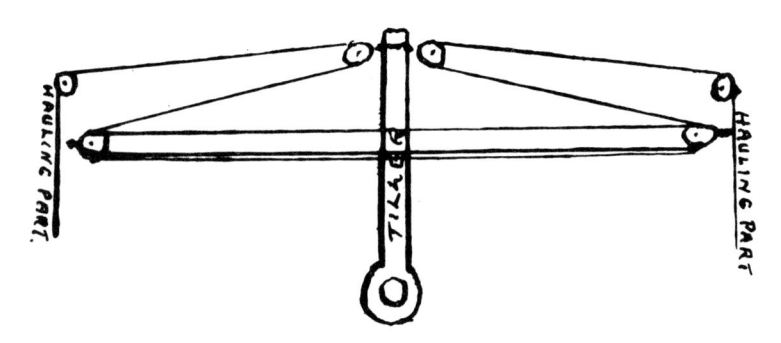

Emergency steering gear for a small vessel.

Secure a double luff tackle to each hauling part of the relieving tackle and take the hauling part of each luff tackle to the winch drum ends. Leading one hauling part over the drum end and the other hauling part under the drum end of a winch. Steer with the winch.

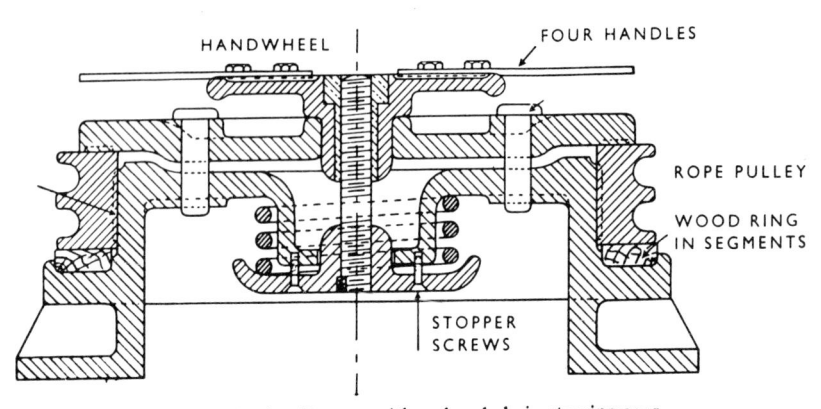

'Dunstos' Rudder Brake, For use with rod and chain steering gear.

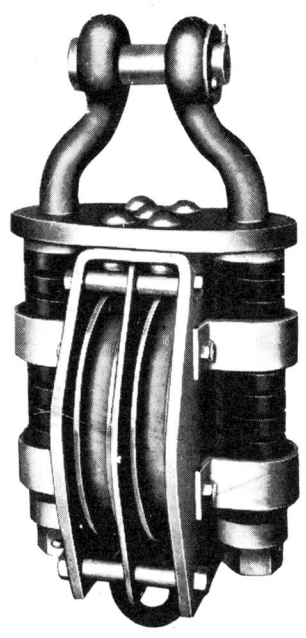

SHOCK ABSORBING BLOCK
FOR EMERGENCY STEERING

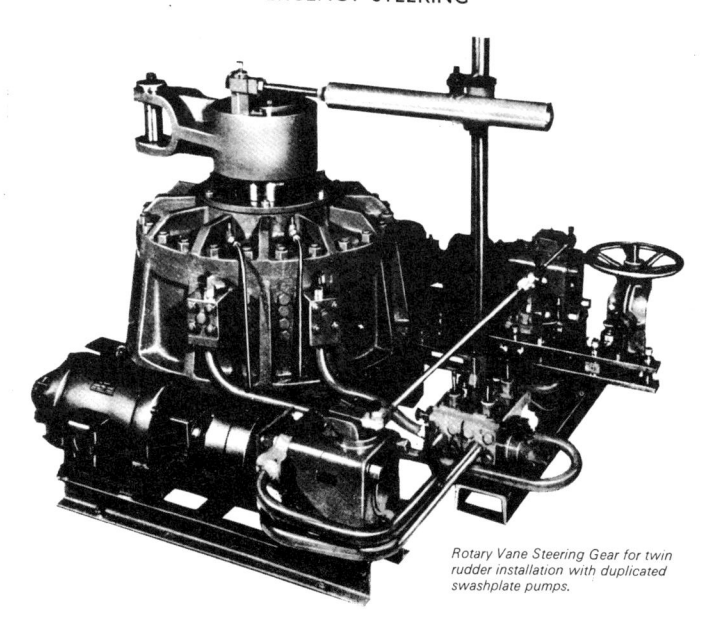

Rotary Vane Steering Gear for twin rudder installation with duplicated swashplate pumps.

Four-cylinder electric hydraulic Steering Gear with duplicated Donkin 'Hele-Shaw' pumps.

Four-ram electric hydraulic Steering Gear with forked tiller and duplicated 'Hele-Shaw' pumps.

HELM ORDERS.

The art of steering a ship can only be gained by practice, basically a good helmsman is an idle man, for the less he turns the wheel, the better the course he will make.

The ship's head, rudder and steering wheel all turn the same way viz., to turn the ship to starboard, turn the wheel to starboard. The angle of turn given to the rudder is shown in degrees on the helm indicator situated forward of the wheel. The maximum amount either way being 35 degrees and with the rudder at this angle the wheel is "Hard-over". With the rudder fore and aft the wheel is "Amidships".

The number of turns of the wheel required to turn the rudder from amidships to hard-over will vary in different ships from about 1/4 turn upwards, the usual being 3½ turns each way. Where only 1/4 turn is required the wheel is normally replaced by a horizontal bar or a half wheel, otherwise it will have eight spokes and in order that it may be easily recognised, one spoke will be marked as the midship spoke.

Due to the effect of the propeller many ships carry a little starboard helm. A good helmsman will watch the ship's head for swing and will normally steer the ship in good weather with the use of as little as two or three spokes. It is essential that the helmsman remains alert at all times, promptly and correctly obeys orders and makes a good course. This applies particularly in narrow crowded waters, where a mistake made by the helmsman could easily result in shipwreck.

All orders received by the helmsman are to be repeated TWICE, once when the order is received and again when the order has been carried out. This is to ensure that the helmsman has both understood and carried out the order correctly. On being relieved a helmsman will give the course to steer and position of the wheel (i.e., amidships, two spokes starboard, etc.) together with any peculiarity such as carrying starboard wheel, to his relief, who will repeat this back to him before taking over. Having been relieved the helmsman will report the course to the Officer of the Watch before leaving the bridge.

A list of typical helm orders might be summed up as follows:—

Order.	Reply by Helmsman.	Action to Take.	Final Report.
Test the steering gear.	Test the gear Sir.	Turn wheel hard over each way and return to amidships.	Gear in order Sir.
Port Ten degrees.	Port ten degrees Sir.	Turn wheel to port until ten degrees is shown on the indicator.	Ten degrees port helm on Sir.

Starboard Twenty degrees.	Starboard twenty degrees Sir.	Turn wheel to starboard until twenty degrees is shown on the indicator.	Twenty degrees starboard helm on Sir.
Hard Aport.	Hard Aport Sir.	Turn wheel as far as it will go to Port.	Wheel hard Aport Sir.
Ease the wheel.	Ease the wheel Sir.	Turn wheel toward midships and stop when five degrees shows on the indicator.	Wheel eased Sir.
Meet her. or Check her.	Meet her Sir or Check her Sir.	Turn wheel till the indicator shows 5 or 10 degrees of rudder against the swing of the ship.	
Midships.	Midships Sir.	Turn wheel to amidships.	Wheel amidships Sir.
Steady.	Steady Sir.	Note compass course and steady ship on that course.	Steady. On course Zero Zero Zero Sir. (or whatever the course may be).
Steer ten degrees to starboard. or Alter course ten degrees to starboard.	Steer ten degrees to starboard Sir. or Alter course ten degrees to starboard Sir.	Give a turn of wheel to starboard, when swinging, ease wheel and steady on new course.	Report new course.
Steer 120 degrees.	Steer 120 degrees Sir.	Alter course to 120 degrees.	120 degrees Sir.
Finished with the wheel.	Finished with the wheel Sir.	Put wheel amidships.	Wheel amidships Sir.

Always alter course to starboard to increase the course number.
Always alter course to port to decrease the course number.
(Except when it is nearer to pass through 000°)

'CHERUB' III SHIP LOG

CHAPTER 5.

THE PATENT LOG AND HAND LEAD LINE

The patent log.
The first log, used in the days of sail, consisted of a specially shaped wood chip attached to a line wound up on a hand reel. This line, after allowing an unmarked length of stray line, was marked first by a piece of bunting and then every 47 feet 3 inches by a piece of fish line. The fish line at the first mark, contained one knot, the second mark two knots, and so on. This line being wound on a hand reel, would run freely when in use. Either a 14 or 28 second sand glass (similar to an egg timer) completed the equipment.

The wood chip was thrown overboard from aft, as the line ran out and the piece of bunting passed the taffrail, the sand glass was turned. When the sand glass ran out, the line was stopped. Using a 28 second sand glass, the number of knots on the last mark to pass the taffrail, would indicate the number of nautical miles the ship was making good through the water in one hour. (When using a 14 second sand glass, the number of knots had to be doubled).

In this manner the word knot came to mean a speed of one nautical mile per hour.

The next step forward was the Patent Log which came with the advent of steam ships. It consists of a rotator towed from the taffrail. The number of turns made in the water by the towed rotator, indicates on a clock attached to the taffrail, the number of nautical miles the ship has travelled through the water.

A Patent Log consists of a shoe fixed to the taffrail. (It is usual to have two shoes, one being fixed on each quarter, so that the log may be streamed on the weather side). A portable clock that is capable of being fitted into either of the shoes and which has a dial at one end and an inglefield clip at the other. A governor consisting of a short length of plaited line with an inglefield clip on one end, that can be clipped onto its counterpart on the log clock, and a spoked wheel at the other end. A ring inglefield clip is fixed to the centre of the fixed wheel on the opposite side to which the length of line is spliced. A log line, consisting of a length of plaited line which may have an inglefield clip on one end. The length of the line is suited to the ship's freeboard and speed, usually somewhere in the region of 40 fathoms (73m). A rotator made of a brass tube to which four vanes have been affixed at an accurate and pre-determined angle. The towing end of the rotator will have a short length of plaited line spliced to it. On the other end of the line is a frog or fish, which is a short hollow brass tube having a large diameter at the centre and a small diameter at each end. Internally, the frog will be ribbed, to prevent the log line when attached, allowing the frog to rotate around the log line.

All the component parts except the shoes, which are fixtures, will normally be kept on the bridge.

To stream the Patent Log. (Set it going).
1. Collect the log clock, governor, log line and the rotator, from the bridge and take them aft. (Take great care not to damage the rotator while carrying it).
2. Fit the clock in the shoe on the weather quarter.
3. Clip the governor to the clock.
4. Flake the log line up and down the deck, to ensure that it will run clear when streamed.

5. Make fast one end of the log line to the governor using a round turn through the ring and a bowline, if an inglefield clip is not attached to the log line for this purpose.
6. Pass the other end of the log line into the hollow end of the frog and out through one of the two large holes in the centre of the frog. Make a figure of eight knot in this end of the log line and force the knot back through the large hole into the centre of the frog.
7. Pay the line overboard from the governor, holding the rotator. When all the line is paid out and forms a clear U astern of the ship, throw the rotator into the water, making sure that it is not damaged by hitting the ship's side as it falls. Alternatively, throw the rotater overboard, let the line run out and ease the strain as the line comes to the clock.
8. When the rotator is properly astern and being towed by the ship, give the governor a twirl to start it off and set the log clock to zero by lifting the glass face and turning the hands anti-clockwise.
9. Plug in the electrical connection to the bridge repeater if one is fitted.
10. Report log streamed and set to the bridge.

The log should always be towed on the weather quarter and handed (hauled in) whenever the ship stops for any reason.

Where no log clock repeater is fitted on the bridge, it is customary for the stand-by man to read the log at eight bells, before going off watch. He then reports the reading and the stern light burning to the Officer of the Watch.

When a reading of the log is required at any other time by the Officer of the Watch it has been customary for the Officer of the Watch to give two blasts on a pea whistle, as a signal to the stand-by man that a reading of the log is required.

To read the patent log, note the position of the large hand on the large dial, this will indicate the number of miles from nought to ninety nine. Of the two small hands on the small dials set into the large dial, the one on the right indicates tenths of a mile and turns anti-clockwise, while the other one on the left indicates hundreds of miles and turns clockwise.

To hand the patent log (haul it in).
Two men should be detailed for this operation.

1. Unplug the electrical connection to the bridge repeater if one is fitted.
2. Stop the governor and haul in a little on the log line by hand.
3. Unclip the inglefield clip or let go the bowline and round turn between the log line and the governor.
4. Continue hauling in the log line by hand and take the end to the other quarter.
5. Pay the free end of the log line out on the other quarter as the rotator is hauled in.
6. As the rotator comes aboard be very careful not to damage it by allowing it to hit the ship's side.
7. With the rotator aboard, haul in the free end of the log line by hand, coiling it left-handed as it comes aboard.
8. Take the rotator off the log line by pushing the figure of eight knot out of the large hole in the frog. Undo the knot and the end of the line can then be withdrawn from the frog.

9. Unship the governor and log clock and take all the gear back to the bridge (being careful not to damage the rotator while carrying it) where the line is to be hung up in a well ventilated but dry position. The clock, governor and rotator are to be stowed in their appointed place.

Note that a log line is always to be hauled in by hand and paid out over the other quarter, to remove the turns that will accumulate in the line as the rotator is being hauled in. Also that it is always to be coiled left-handed.

Many ships at one time streamed the patent log from a boom amidships but the fittings required to stream a log in this manner, are no longer being manufactured. To hand one, cast a small grapnel on a line out over the log line and haul the log line and rotator aboard.

For that matter the patent log is rapidly falling into disuse. Modern ships are now fitted with an underwater log which does not require any attention from deck ratings.

The hand lead line.

The hand lead line consists of about 25 fathoms (45.7m) of $1\frac{1}{8}$ inch circumference (9mm) cable laid hemp line, marked at various depths and having a 7 or 14 lb. (3 or 6kg) lead attached to one end. It is used while the ship is under way in narrow waters, to ascertain the depth of water.

In this modern day and age, it is seldom used, except in pilotage waters, when it is usual for the pilot to bring his own leadsman aboard. Even so, it can be a most useful asset in an emegency and all deck personnel should have some idea of how to use it.

To make a hand lead line, break out a hank of hand lead line. Put an eye splice in one end and a whipping on the other. Stretch and soak the line by towing it astern. Mark off 2, 3 and 5 fathom lengths on a deck plank with chalk. Haul in the line and while it is still wet, commencing at the end of the splice (the length of the lead, known as "The benefit of the lead" is not included in the measurement) mark off on the line, the various places at which it is to be marked.

Mark the line as follows:—

At   2 fathoms a piece of leather with two tails.
At   3   "   "   "   "   "   three tails.
At   5   "   "   "   "   "   white linen.
At   7   "   "   "   "   "   red bunting.
At 10   "   "   "   "   "   leather with a hole in it.
At 13   "   "   "   "   "   blue serge.
At 15   "   "   "   "   "   white linen.
At 17   "   "   "   "   "   red bunting.
At 20   "   "   "   "   "   cord with two knots in it.

Pass the eye in the end of the lead line, through the leather thong in the hand lead, pass the lead through the eye and draw tight. Coil up left-handed and hang in a dry but well ventilated place on the bridge.

To take a cast of the lead (find the depth of water) in sailing ships, the leadsman would go outside the bulwarks on the weather side and stand on a baulk or plate to which chains holding the shrouds were shackled and was said to be "In the chains".

In steamers the leadsman stands on a small hinged platform let down from the bulwarks in the vicinity of the bridge, on the side on which the shallowest water is to be expected. The platform will have a canvas apron round its perimiter, to protect the leadsman from water. This platform is still called the "Chains".

A leadsman can only become proficient by practice. For ship speeds up to about 6 or 7 knots, good soundings can be obtained with a 7lb. (3kg) lead. For speeds up to 10 knots a 14 lb. (6 kg) lead is necessary together with a well experienced leadsman. Over 10 knots hand sounding is not practical.

To take a cast of the lead, make fast the end of the lead line to a strong point. Stand in the chains with the coil of line in the left hand. Holding the lead line in the right hand, about 2 fathoms (3.6m) from the lead, either by a toggle placed in the line, or by taking a turn of line around the palm and thumb. Swing the lead back and fore, some leadsmen swing the lead a couple of times in a full circle over their head (probably more for show than efficiency). Let go, allowing the lead to fly as far forward as you can make it go. As the line runs off the left hand, use the right hand to allow the line to pass through and feel for bottom as the lead comes up and down. Ascertain the depth by the marks and call it out in a clear voice. Haul in the lead, recoiling it at the same time in the left hand, ready for the next cast.

The various marks were made of different materials, so that on a cold dark night, the leadsman could place a mark against his lips and know by the feel of it, which mark it was. Today, of course, the leadsman is able to use an electric light placed in a strategic position but the old traditional method of marking is still used.

When calling the lead, the marked fathoms, of which there are nine, 2, 3, 5, 7, 10, 13, 15, 17 and 20 are called "Marks". While the unmarked fathoms, of which there are eleven, 1, 4, 6, 8, 9, 11, 12, 14, 16, 18 and 19 are called "Deeps". Soundings are called to quarter fathoms but the words "three quarters" are never used.

The calls should be given as follows:—

At nine fathoms	"Deep nine".
At nine and a quarter fathoms.	"And a quarter nine".
At nine and a half fathoms.	"And a half nine".
At nine and three quarter fathoms.	"A quarter less ten".
At ten fathoms.	"By the mark ten".
When no bottom is found.	"No bottom at XYZ fathoms".

The number of fathoms always being to the nearest fathom and the last word of the call.

The hand lead line is also sometimes used when coming to anchor, to ascertain when the ship has stopped moving over the ground, for this is not always apparent in a tideway. To do this, the leadsman drops the lead overside and leaves it on the bottom, then notes which way the line leads, calling out accordingly, with the line leading astern. "Ship going ahead Sir", "Ship moving astern Sir", with the line leading ahead or "Ship stopped Sir", when the line remains up and down.

The deep sea hand lead line.

The deep sea hand lead line is seldom, if ever, used today. However, it is still required to be carried on passenger ships and a knowledge of it is therefore required.

A deep sea lead weighs 28lb. (12.7k) and has a depression in its base, into which a mixture of white lead and tallow (2 of tallow to 1 of white lead) is pressed. This is called "Arming the lead", its purpose being to collect a little of the sea bed and so learn the nature of the sea bed (mud, sand, etc.) The nature of the sea bed, coupled with the depth of water can, on consulting the relevant chart, often give a very good indication of the ship's position to the navigator.

The deep sea hand lead line is made of 1½" circumference (12mm) cable laid hemp approximately 120 fathoms (220m) long and marked to 100 fathoms. It is spliced, stretched, soaked, measured and marked in exactly the same manner as the hand lead line. Beyond 20 fathoms (36.5m) it is marked at every ten fathoms (18.2m) with a piece of cord containing one knot for every 10 fathoms (18.2m). i.e., 3 knots at 30 fathoms, 4 knots at 40 fathoms and so on, and at every intervening 5 fathoms (9.1m) with a single knot. It is then wound on a hand reel and kept in a dry but well ventilated place.

To take a cast of the lead with the deep sea hand lead line, the ship should first be stopped. The lead is armed and taken forward to the weather side of the fo'castle head, with some spare line. The line is then led outside the ship to the leadsman standing in the weather chains. The man on the fo'castle head will swing and let go the lead, shouting to the leadsman "Watch there watch" as he does so. The leadsman will then attempt to find bottom. Several casts, giving "No bottom" may be required before the leadsman is able to determine the approximate amount of line to let run out, to take a good cast.

When the cast has been taken and the lead hauled in, the arming is to be cut out and taken to the bridge, for inspection by the navigating officer.

The deep sea hand lead line was replaced many years ago by the Patent Deep Sea Sounding Machine, which in it's turn has been superseded by the echo sounding machine. Since the Patent Deep Sea Sounding Machine is no longer in regular use, it has been dropped from the latest E.D.H. and A.B. examination syllabus and therefore is not included in this text book. Likewise the echo sounding machine is a scientific instrument, the care, operation and maintenance of which is a specialists work and therefore beyond the scope of this book.

THE MARKING OF A METRIC HAND LEAD LINE.
1, 11 and 21 metres. One strip of leather.
2, 12 and 22 metres. Two strips of leather.
3, 13 and 33 metres. Blue bunting.
4, 14 and 24 metres. Green and white bunting.
5, 15 and 25 metres. White bunting.
6, 16 and 26 metres. Green bunting.
7, 17 and 27 metres. Red bunting.
8, 18 and 28 metres. Blue and white bunting.
9, 19 and 29 metres. Red and white bunting.
10 metres. Leather with a hole in it.
20 metres. Leather with a hole in it and two strips of leather.
30 metres. Leather with a hole in it and three strips of leather.
40 metres. Leather with a hole in it and four strips of leather.
50 metres. Leather with a hole in it and five strips of leather.
All 0.2 metre markings. A piece of mackeral line.

6 feet = 1 fathom. = 1.8288 metres. - - - - - - - - - 1 metre = 3.2808 feet.

The above metric hand lead line is that introduced by the Hydrographer of the Navy for hydrographic survey. It is possible that a simplified version will be introduced for use in merchant shipping.

CHAPTER 6.

ANCHORS, CABLES AND WINDLASS.

A ship will normally carry four anchors, three bower and one stream anchor. On modern ships, stockless anchors are used as bower anchors. The shank is hinged and when the anchor is let go, the weight of the cable pulls the shank over and the flukes dig into the ground. The main advantage lies in the fact that having no stock, the anchor can be hove home in the hawse pipe and remain there while the vessel is on passage. A stock anchor, when it is let go, is pulled over by the weight of the cable, the stock rests on the ground and tips the anchor so that one fluke bites the ground. When it is weighed, it has to be "catted", that is to say that it has to be hove up onto the fo'castle head by means of a cat davit, to be stowed.

There is a bower anchor on each bow and the third, which is a spare, will be stowed in an accessible position on the fore deck.

The stream anchor is for use from the stern. If it is on a ship regularly trading to a port where stream anchors are required, it will probably be a stockless anchor, fitted in a stern hawse pipe. Normally, however, stream anchors are seldom used, except in an emergency, and a small stock anchor is stowed in an accessible position on the after deck, from where it can be hoisted outboard by a derrick, if the need to use it arises. Under these circumstances, a strong wire rope will normally be used for a cable.

The two bow anchors will be fitted with a total of between 201 and 330 fathoms (384 & 604m) of chain cable, according to the length of the ship. Half the cable (7 to 11 shackles) being on each anchor. Each link in the cable is studded, this not only adds to the strength but helps to stop the cable kinking. The cable which is in 15 fathom (27.4m) lengths, is joined up by shackles. Each shackle being plainly marked.

Apart from the bending on shackle, which is a lugged shackle joining the shank of the anchor, to the cable. Lugged shackles when used as joining shackles, will be fitted so that the rounded part of the shackle is forward, this prevents the lugs catching on anything, when the anchor is let go. A shackle known as the ganger shackle is always placed between the windlass and the hawse pipe. It's purpose is to allow the cable to be broken from the anchor, with a minimum of trouble. The end of the cable can then be led to an eye in the bow and passed out for securing to a buoy.

At the end of the first 15 fathom (27.4m) length of cable, is the first shackle. The first *studded* link on each side of the shackle will have turns of seizing wire wound round the stud. These links together with any open links and the shackle should all be painted white. This method of marking applies at each shackle, so that at the sixth shackle, the sixth studded link on each side of the shackle will be marked with seizing wire wound round the stud and the length of cable between the two markings, over the shackle, should be painted white. With this method of marking, it is a simple matter when the anchor is dropped, to sight shackles as they go over the gipsy and brake the windlass as a shackle crosses the gipsy to ascertain the number of shackles in the water.

When lugged shackles are fitted, there is an unstudded or open link, each side of the shackle. These open links are not counted when marking the shackles.

When Kenter shackles are fitted, all the links are studded and therefore counted.

BOWER (STOCKLESS) ANCHOR

- Bending on shackle
- Shank
- Pea or Bill
- Fluke
- Arm
- Tripping Palm
- Crown

Chain hook

Devil's Claw

STOCK ANCHOR

- Bending on shackle
- Stock
- Key
- Stowed position of stock
- Shank
- Gravity band
- Pea or Bill
- Fluke
- Arm
- Crown

The gravity band is for lifting from stowed position

KENTER SHACKLE

Pin

Stud pin

Lead pellet

LUGGED SHACKLE

Pin held in position with wood spile

Pin

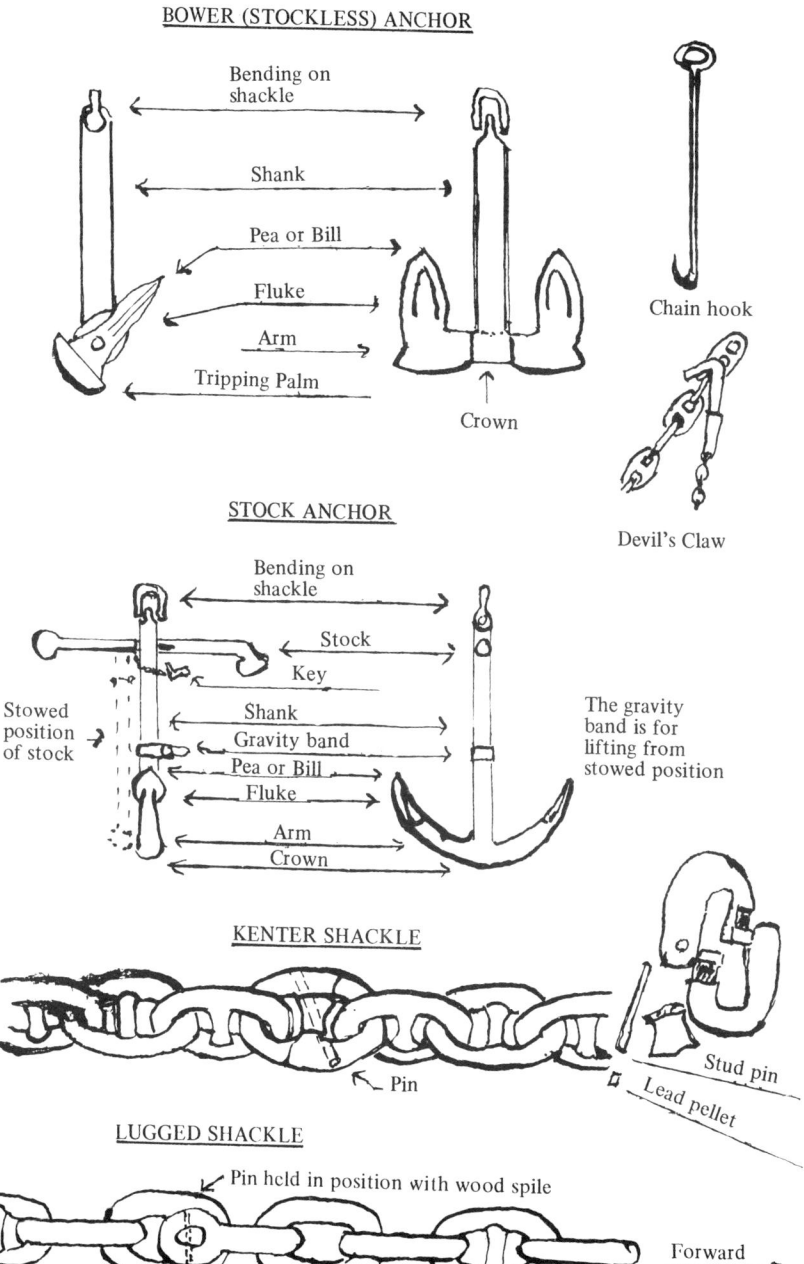

Open links

Forward →

Donkin Electric Windlass

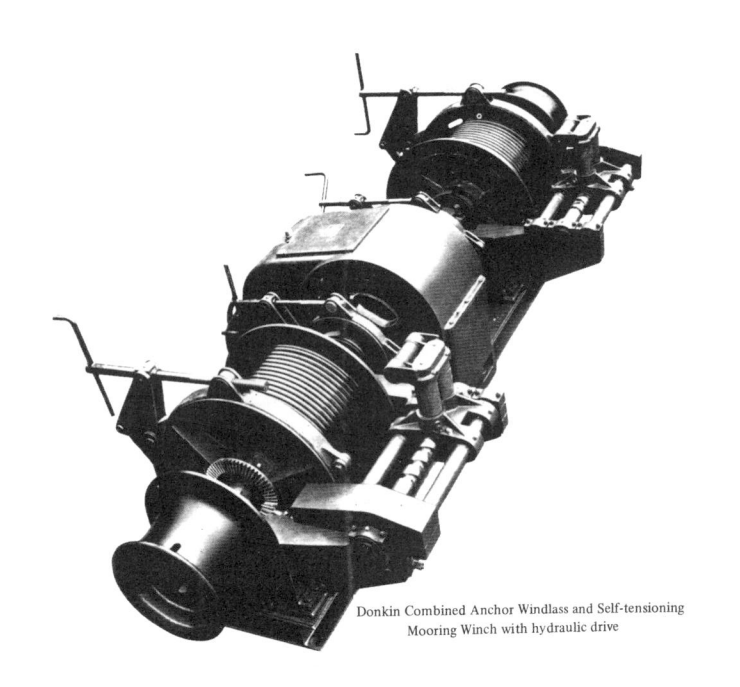

Donkin Combined Anchor Windlass and Self-tensioning
Mooring Winch with hydraulic drive

When the anchor is being let go or weighed (lifted), as each shackle comes over the gipsy, a number of strokes, corresponding to the number of the shackle, should be made on the ship's bell.

The end of each cable is made fast in the chain locker with a lashing.

Lugged shackles are a D shackle having a pin through the ends of the U, that does not protrude beyond the width of the shackle. This pin is kept in position by a hardwood pin called a spile, going through holes in both the shackle and pin.

Kenter shackles are really a split link, which is in four parts, two half-links, a stud and a metal pin or spile that is kept in place by having a lead plug hammered in over the top.

When the anchor is being weighed, a man (with a light if necessary) is required to go down to, but NOT INTO, the chain locker. Chain lockers today, are termed self-stowing. This does not prevent the chain piling as it comes in. An attendant is necessary to prevent the piles rising too high and toppling over, so burying the top chain underneath. The locker man is required to pull the chain occasionally, to let it slide down the side of the pile, before the pile is high enough to fall. This is best achieved from OUTSIDE the locker by passing a bight of rope around the chain. If the chain fouls the rope, let go one end and pull the rope out. ON NO ACCOUNT IS A CHAIN HOOK TO BE USED. A chain hook will foul in the chain and be lost. If anything like this happens STOP THE WINDLASS. The cable must then be paid out with the windlass in gear and the hook or whatever it is, recovered. A chain hook, or anything else, left in a locker, will spell almost certain death to someone, when the anchor is let go and it flies up out of the spurling pipe.

Chain hooks are for use when handling the cable on deck and NOT for stowing cables in lockers.

The windlass.

The windlass is a machine primarily provided as a means of letting the anchor cable run out and hauling it in again.

It may be steam, electric or hydraulic and will normally have two gipsies, (or cable lifters), which can be put in and out of gear, both together and separately. A brake for each gipsy and reversing gear. It usually has a winch drum end on either end of the main shaft, provided for the purpose of hauling on mooring lines when tying up, letting go or warping. A small spur, called a kicking strap, is fitted under each gipsy just forward of the spurling pipe, for the purpose of knocking out any links of the cable that get caught in the gipsy and ensuring that they go down the spurling pipe, as the cable is hove in. A compressor, (or bowstopper), which is a device that can be used to hold the cable and so take the strain of the cable off the windlass brake, is placed between the gipsy and the hawse pipe.

Getting ready to anchor.
1. Request the engine room for power on deck.
2. Take to the fo'castle head a spike, hammer or crowbar, oil-can, goggles and at night a torch. Take off the hawse pipe covers. Let go the lashing in the chain locker.
3. Make sure the windlass is out of gear and that the brakes are on.
4. Turn the windlass over slowly and oil the moving parts. On a steam windlass the drain cocks must all be opened and the water allowed to drain off before the cocks are closed again.

5. Put one anchor in gear. (see that gears are clear to engage first).
6. Remove the devil's claw and any other lashings, the compressor bar and the cement or other filling from the spurling pipe.
7. Make sure that the weight of the anchor is held by the brake and that the gears although in, are clear. Take out of gear.
8. Prepare the other anchor using the same routine. (5,6 & 7)
9. Inform the Officer concerned that the anchors are ready for lowering clear of the hawse pipe.
10. Place one anchor in gear.
11. When ordered to lower away by the Officer. Take off brake and lower slowly until the anchor is out of the hawse pipe.
12. Screw brake tightly home and take out of gear.
13. Repeat 10, 11 and 12 with the other anchor if required. Return the gear.

Letting go the anchor.
Wear goggles. Let go the brake when ordered and brake as required after the anchor has hit the bottom. One man to strike the bell as the shackles go out. 1 Shackle, 1 bell. 6 Shackles, 6 bells. etc. Put on the compressor bar and screw the brake tightly home, when sufficient cable has been paid out. Hoist the anchor ball or anchor lights. Return the gear.

Weighing the anchor.
1. Take to the fo'castle head a hammer or crowbar, hose, oil-can and at night a torch. Couple up the hose, lead it to the hawse pipe and open the cock.
2. Request the engine room for power and water on deck.
3. Make sure the windlass is out of gear and the brakes are on.
4. Turn the windlass over slowly and oil the moving parts. On a steam windlass the drain cocks must all be opened and the water allowed to drain off before the cocks are closed again.
5. Put the anchor in gear (see that the gears are clear to engage first).
6. Send a man to the locker (with a light if necessary).
7. Remove the compressor bar and when ordered to do so, take off the brake and commence heaving in the cable. See that the hose is running and have a man to wash the cable as it comes in. Inform the man in the locker that you are about to heave away. Another man should stand by the bell to ring the shackles as they come in.
8. When the anchor is hove home, inform the man in the locker. Apply the brake tightly and the compressor bar. Ease the gears and take the windlass out of gear. Stop the water, close the cock and uncouple the hose. Take down the anchor ball or anchor lights and ring the bell rapidly when the anchor is aweigh. Return the gear.

Securing the anchor for sea.
1. Ensure that the brake is tightly home, the compressor bar on and the windlass is out of gear.
2. Place the devil's claws on the cables and screw the bottle screws up tightly.
3. Put the hawse pipe covers in position and if proper plates are supplied for the spurling pipes, place them in position and cover with the canvas coat. If there are no proper plates for the spurling pipes, the cables must be well wrapped with burlap or sacking in the spurling pipe. The pipes are then to be filled over the burlap with a good thickness of cement (sufficient to

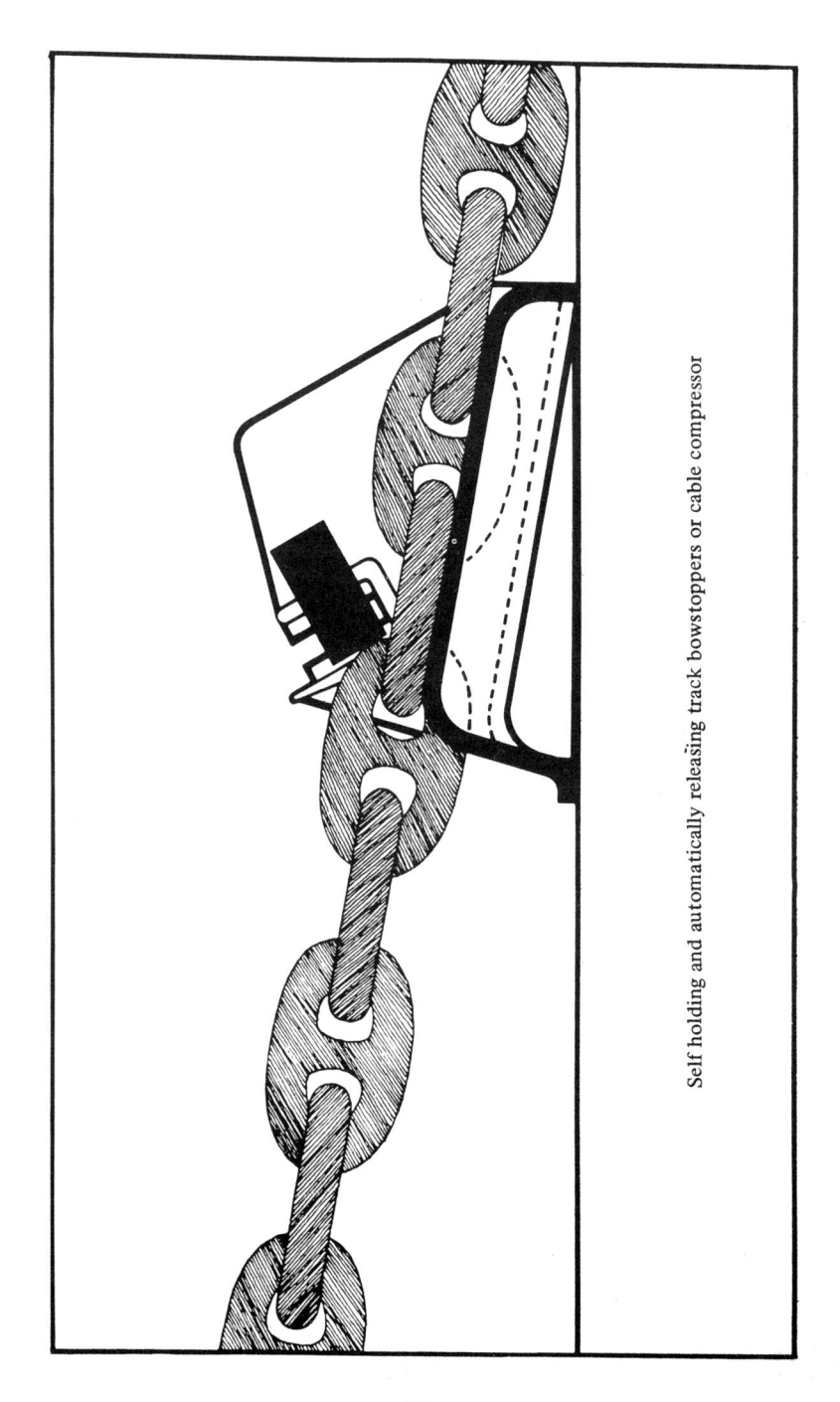

Self holding and automatically releasing track bowstoppers or cable compressor

withstand any movement of the cables or weather damage). To help prevent movement, lash the two cables together beneath the spurling pipes in the chain locker.
4. Return all gear.

To break a lugged shackle, punch out the pin with a punch and munday hammer, the wood spile will break. To break a Kenter shackle, punch out the pin.

ACCIDENT PREVENTION'
Be sure the brake is tightly home and the gears eased before taking the windlass out of gear.
Do not leave the windlass in gear.
Wear goggles when letting go the anchor.
Do not go in the chain locker to stow the cable.
Do not use chain hooks in the chain locker.
See that the spurling pipes are made properly watertight.
When it is necessary to let go an anchor or pay out cable and a man is in the locker (as might happen when mooring) The order "Stand clear in the locker" must be given, replied to and carried out before the anchor or cable is allowed to run out.
Extracted from— **Merchant Shipping Notice No.622**

MISCELLANEOUS OPENINGS IN FREEBOARD AND SUPERSTRUCTURE DECKS—SPURLING PIPES.
1. The Department of Trade and Industry wish to draw attention to the following two incidents where the ingress of water through spurling pipes was a contributory cause to these ships becoming casualties. In the first case, referred to below, the ship foundered resulting in a loss of life.
2. A loaded ship on passage from Holland met with heavy weather in the English Channel. The canvas covers to the spurling pipes were torn away and the chain locker and forecastle store became flooded. The ship being already down by the head the additional flooding was sufficient to bring the well deck under water thereby placing the air pipes and subsequently the cargo hatchways in jeopardy. Progressive flooding occurred and eventually the ship was lost.
3. The other ship also outward bound from Holland encountered heavy seas which were sufficient to break the cement which plugged the starboard spurling pipe. Sea water thus gained access to the chain locker and the forward stores space producing trim by the head and bringing the foredeck awash. Fortunately the Master took early action and returned to port where the spurling pipes were re-cemented.
4. In the first case the mere fitting of canvas around the spurling pipes was insufficient but had the canvas been supported by close fitting steel plates a more efficient means of preventing the ingress of water would have been provided. The second case showed that the quantity of cement used must be adequate not only to plug the spurling pipes but also to prevent lateral movement of the cables within them.
October 1971.

CHAPTER 7.
FIREFIGHTING AND LIFESAVING APPLIANCES.

FIRE PREVENTION.

Three things are required to produce and sustain a fire. Heat, fuel and oxygen, an absence of any one of these three will cause the fire to go out. On the other hand, heat, provided there is sufficient of it will cause a fire without the necessity of having to light it.

The amount of heat required to start a fire will depend entirely on the temperature at which the fuel will ignite spontaneously. With some substances such as fuel vapour this is very low, while with others it can be exceedingly high, anything and everything will burn. The reason some substances appear not to burn lies in the fact that the heat produced in most fires is insufficient to ignite them.

Obviously the most successful method of fire fighting is fire prevention. Since fuel and oxygen are always present, this is best achieved by eliminating the heat.

Spontaneous combustion (setting light to itself) may be caused in a number of cargoes like coal or cotton by damp. Water in some substances, if there is no ventilation to carry the heat away, will slowly but surely (as it does in a haystack or manure heap) increase the temperature, until the small amount of oxygen in the fuel ignites and the fuel smoulders. As soon as a large volume of air (which contains oxygen) comes in contact with the smouldering fuel (as it will when the hatches are removed from a hold) there is a fierce fire. Cargoes which are liable to spontaneous combustion should therefore be kept well ventilated, to remove any heat before it builds up to the point where the cargo will ignite.

Bedding and ship's stores should never be stowed on top of, or too close to any type of radiator. Neither should a store room be filled to capacity so that the air space around an electric light bulb either in the deck head or anywhere else becomes restricted. Electric light bulbs and all radiators become very hot when there is no circulation of air to carry the heat away and will quickly set fire to any bedding or stores in the immediate vicinity, or to a towel or shade placed too close to it. Do remember to switch the light out before leaving any cabin or storeroom.

Defective electric wiring is liable to heat or spark and set fire to anything it touches or to any inflammable vapour. This is a very real risk when portable leads are connected to various electrical equipment. All electrical equipment and leads should be examined by, and permission to use them obtained from the ship's electrician or other competent person. Above all, disconnect all electrical equipment, particularly any electric iron or fire if one has been in use, before leaving a cabin. Also close the ports, forced ventilation and the door if you expect to be away any length of time.

Matches must always be extinguished and cigarette ends stubbed out, before being discarded into a proper receptacle. Do not throw match or cigarette ends about. NEVER smoke in bed, should you fall asleep and the cigarette set light to the bed clothes, there is no guarantee that the smoke will wake you.

In the accommodation of any ship a fire will always spread rapidly along an alleyway or air duct if given a chance (paint in the alleyway and dust in the airduct will provide ample fuel). Doors, ports and ventilating trunks in all empty cabins should be kept shut, to help prevent the spread of any fire that may start in the cabin. In all unoccupied cabins the lights and radiators should be turned off. In the event of a fire in a cabin. Raise the alarm. The door should not be opened until hoses have been rigged and water is available.

Do not leave an aerosol in strong sunlight or near a source of heat. Do not allow the spray from any aerosol to come in contact with a naked flame or heated surface, the contents are highly inflammable.

Avoid collections of either damp or greasy waste or cloth in cabins, storerooms or workshops. This sort of thing is very liable to spontaneous combustion.

Never smoke in a no-smoking area.

FIRE FIGHTING.
The most important aspect of fire fighting is immediate action. To sound the alarm and then either attempt to put it out or restrict the amount of air available for the fire to burn and so prevent it spreading.

There are two methods of extinguishing a fire:—
1. By removing the heat. This can usually be accomplished by playing water (or sometimes in the case of a fire in a cargo, steam) onto the source of the fire, and keeping the surrounding area cool to prevent the fire spreading.
2. By excluding oxygen. This is achieved by closing all doors and ventilators and replacing the air in the vicinity of the fire with an inert gas such as carbon-dioxide (CO_2) or by blanketing the fire with a powder, foam or blanket.

Fires in the proximity of electrical wiring will require the extinguishing agent to be a non-conductor of electricity. NEVER use water or foam near electrical equipment or wiring without first pulling the main switches or fuses and ensuring that the equipment is not live. Electricity travels through water and can kill you instantaneously.

The first line of defence against a small fire is provided by means of portable fire extinguishers. These are for use before the fire has had a chance to obtain a firm hold. There are several different types and each has its own particular uses, advantages and disadvantages.

Get to know the exact location of all the fire alarm points and fire fighting implements and portable fire extinguishers in the ship, as soon after joining as you reasonably can. Read and memorise the instructions on all the portable fire extinguishers. The middle of the night is no time to be searching for extinguishers and trying to read the instructions.

Do not use sand from the fire buckets to scrub the deck or for anything else except fighting fire.

Do not use a water fire bucket for any purpose whatsoever except fighting fire. Whenever you pass a water fire bucket with the hose, replenish the water in the bucket.

Do not play games with portable fire extinguishers or they will be empty when you really need them.

Do not use fire hoses for washing down the decks.

On fire drills, learn the correct way to hold a fire nozzle when the water pressure is full on. Keep the hose under the armpit and place your hands palm down on top of the nozzle so that you can push it down. A fire hose that takes charge always tries to rise, control is kept by holding the nozzle down. If a nozzle takes charge, open another cock to reduce the pressure, before attempting to control the nozzle. Learn also how to adjust and operate the breathing apparatus on board and learn the signals off by heart. You will not be able to read them in the dark and smoke.

If a person's clothing catches fire, remember that flames burn up. Throw the person to the ground before attempting to smother the flames.

All used or partially used portable extinguishers are to be recharged at the first practical opportunity.

Portable Fire Extinguishers

FIRE CLASS ACCORDING TO BS. 4547 1970	EXTINGUISHING PRINCIPLES	A.B.C. ALL PURPOSE POWDER	STANDARD & SAPPHIRE DRY POWDER	METAL POWDER	CO_2 GAS	FOAM	WATER	B.C.F.
A Fires involving solid materials usually of organic nature in which combustion normally takes place with the formation of glowing embers. Wood, paper, textiles etc.	Water Cooling or Combustion Inhibition	Yes Excellent — Rapid flame knockdown and excellent protection against re-ignition.	NO — Will control small surface fires only.	NO	NO — Will control small surface fires only.	YES — Has smothering, cooling and sealing action.	Yes Excellent — Good penetration and rapid cooling of combustibles below fire point prevents re-ignition.	YES — Rapid flame knock-down.
B Fires involving liquids or liquefiable solids. Burning liquids oil, fat, paint etc.	Flame Inhibiting or Surface blanketing and cooling.	Yes Excellent — Screen of dry powder shields operator. Rapid Flame knockdown.	Yes Excellent — Screen of dry powder shields operator.	NO	YES — Leaves no residue. Does not contaminate food.	Yes Excellent — Foam blanket gives protection against re-ignition and cools the liquid fuel	NO — Water will spread the fire.	YES — Provides quick flame knock-down.
C Fires involving gases.	Flame Inhibiting.	YES	YES	NO	YES	YES	NO	YES
D Fires involving metals, Magnesium Sodium Titanium Zirconium	Exclusion of oxygen and cooling.	NO — Use of wrong medium would cause explosion.	NO	Yes Excellent — Forms a crust over burning metal and excludes oxygen.	NO — Use of wrong medium would cause explosion.	NO	NO	NO
FIRES INVOLVING ELECTRICAL HAZARDS	Flame Inhibiting	YES — Non-conductor.	YES — Non-conductor.	NO	Yes Excellent — Non-conductor Leaves no residue.	NO — Foam is a conductor.	NO — Water is a conductor.	YES — Non-conductor.
PRESSURE SOURCE		CO_2 Cartridge	CO_2 Cartridge	CO_2 Cartridge	Gas compressed in the cylinder.	Chemical reaction or CO_2 cartridge.	Water/CO_2–CO_2 Cartridge Soda Acid-Chemical reaction Aircharge–Stored pressure.	Nitrogen
SIZES AVAILABLE		3 kg. 6 9 12	1 kg. 75 kg. 1·5 2 3 6 9 12 50 lb 22·7 kg.	9 kg.	1·14 kg 2·27 3·2 4·5 6·8 22·7 kg. 9·1 13·6 18·1	4·5 litre 10 80 136	10 litre	·68 kg 1 kg 1·36 kg
RECHARGING		On site	On site	On site	By arrangement with supplier.	On site	On site	Replacement Cylinder on site.

1 lb = ·454 kg   1 Gall = 4·544 litres

FIRE SMOTHERING BLANKET

Model PMF2
Approved by the Fire Offices Committee
(a)

Portable foam fire Extinguisher (a)

TO REFILL: Pour in cold water to inside level, add the foam concentrate. Screw on CO_2 cartridge to head cap, and tighten to cylinder. (Note – mixing is not required).

80

Fire buckets	Not to be used on oil or electric fires. Throw the water onto the seat of the fire.
Sand buckets.	Cover spilt oil with sand to prevent the fire spreading.
Asbestos blankets.	Placed in the galley near the fish fryer. Cover a pan of burning fat with a blanket.
Blankets.	(Wet and wool if possible) Suitable for smothering a small fire. If wet, do not use on electrical equipment. (Wool singes but does not easily burn).

Carbon tetra-chloride (C.T.C.) extinguishers give off toxic vapours and are not therefore approved for use aboard ship.

MAJOR FIRE APPLIANCES.

Hoses. Fire hoses and hydrants are placed at strategic points. Fire hoses that are not permanently stowed on reels, should be made up in "Dutch rolls". To make a dutch roll:— Stretch the hose and double it, laying one half of the hose upon the other, roll from the centre so that both couplings will be on the outside of the roll. A hose rolled is this manner is much more easily stretched in an emergency, than a hose rolled from one end.

To fight a fire of the Class A type in the accommodation, use a jet nozzle, playing the jet on the seat of the fire. Other hoses should if possible be rigged and played on the bulkheads, decks and deck-heads, to prevent the heat from the fire spreading.

To fight a fire of the Class B type, two methods are available.
1. Using a spray nozzle allow a fine mist spray to fall on the seat of the fire. The larger the area the curtain of spray can be made to cover, the better.
2. Using a foam making branch pipe coupled to a low expansion foam making compound and with suitable water pressure, adjusted by means of the cock to suit. Allow the foam to fall lightly on the far side of the fire and gradually cover the fire. When feasible the foam should be directed onto a bulkhead and allowed to flow down the bulkhead onto the fire. If foam is allowed to hit the oil with any force, it will sink, become oil covered and useless. It will also splash the oil and so cause the fire to spread. Where the oil on fire is in a raised tank, buckets of sand set beneath any drips, will help to prevent the fire spreading.

Unless the bilges are clear and the bilge pumps working, so that excess water can be pumped out. Large quantities of water should not be pumped into a hold, tank or engine room, because an excess of free surface water may cause the ship to lose stability and possibly capsize. Fortunately a large quantity of foam can be made with very little water.

Steam. Some ships are equipped with remotely controlled steam cocks situated in the holds, others are supplied with lengths of steam hose which can be coupled to the deck steam line, for dealing with fires in compartments. Provided all ventilators are effectively sealed, steam can often contain a fire in many types of cargo, by preventing a heat build up and the consequent spread of the fire and may succeed in actually dowsing the fire.

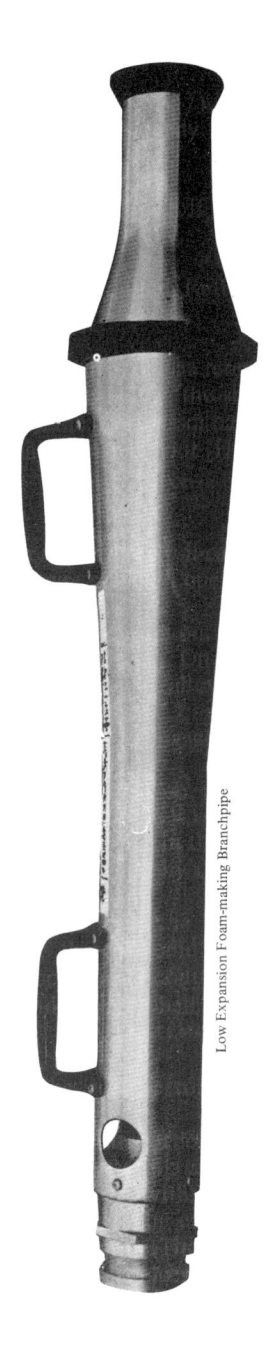

Low Expansion Foam-making Branchpipe

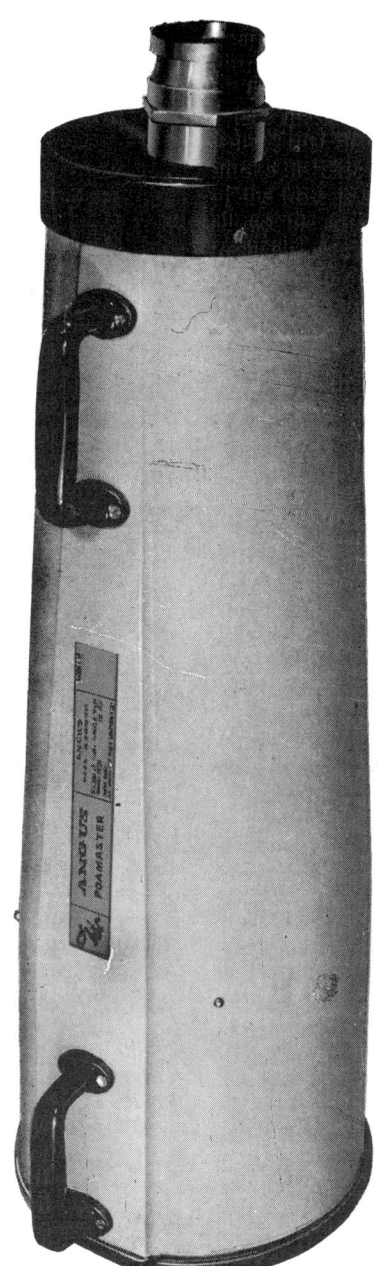

Medium Expansion Foam-making Branchpipe

Showing generator in use with canvas trunking to carry the foam

ANGUS TURBEX WATER
TURBINE

HIGH EXPANSION FOAM
GENERATOR

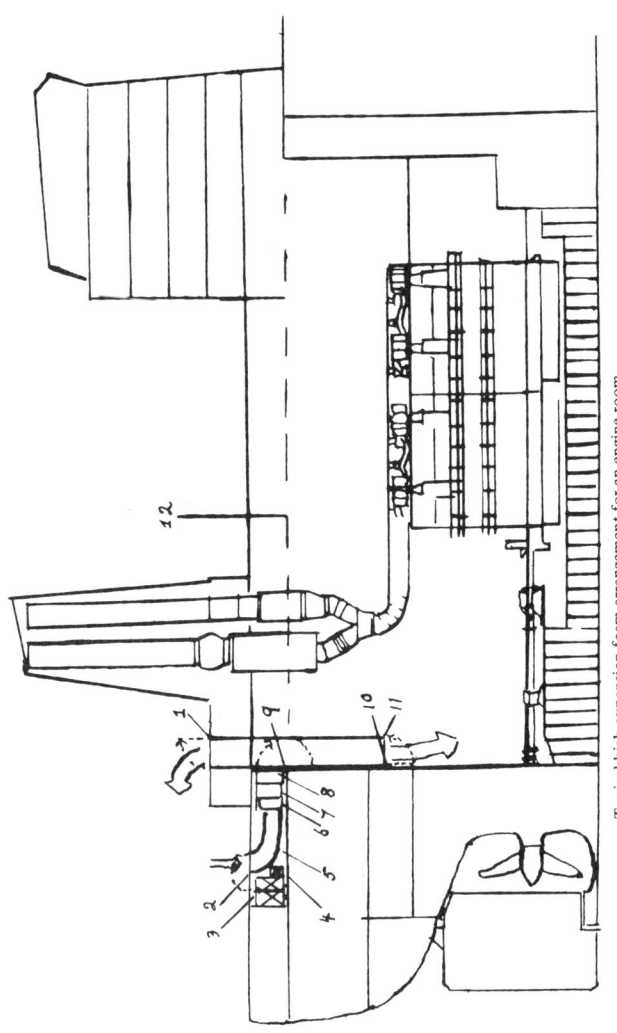

Typical high expansion foam arrangement for an engine room.

1. Test outlet.
2. Air inlet.
3. Water storage.
4. Foam compound storage.
5. Air duct
6. Inspection door.
7. Motor fans and pumps.
8. Foam production section.
9. Foam outlet.
10. Cover. (closed on test).
11. Foam outlet.
12. Engine room.

The emergency fire pump is connected to the water storage tanks, to ensure a continuous water supply.

Foam. Three types of foam are available for fire fighting:—

Low expansion foam. Which consists of small bubbles made of water containing a percentage of an animal offal based foam compound and air. This type of foam may be thrown in a jet for short distances and is suitable for use on burning liquids. Care must be taken not to allow the jet to splash the liquid and so spread the fire.

When fighting an alcohol type fire, a specially manufactured type of foam compound must be used (not suitable for use with salt water).

Medium expansion foam. Consists of somewhat larger bubbles made of water containing a percentage of detergent and air. This type of foam may be thrown for very short distances from a foam making branch pipe or be poured from a foam generator. It is suitable for use in a compartment, where it will be contained, against class A and B fires.

High expansion foam. Contains still larger bubbles, it cannot be thrown but has to be manufactured in a foam making generator and poured onto the fire. The size of the bubbles depends on the size of the mesh in the generator, through which the water and detergent foam compound are blown. It is most suitable for fires in compartments which have considerable empty spaces (such as an engine room), as the foam will fill the empty spaces, thus depriving the fire of the oxygen they would otherwise contain. Moreover there is very little water damage resulting from its use.

Unlike conventional fire fighting methods, no attempt at restricting ventilation should be made when using expanded foam. The currents of air feeding the fire, will carry the foam bubbles to the seat of the fire in a most astonishing manner. Bubbles burst with the heat of the fire, turn to steam and help to both cool and smother the fire, gradually bubbles will encroach on the fire and extinguish it. Meanwhile, the detergent in the foam cleans the air of smoke and soot, making it possible for the operators to obtain a good view of the results of their efforts.

Difficulty arises however in dispersing the foam, for the bubbles may have a considerable length of life. If possible it can be removed by suction, or be swept aside with boards. An alternative method is to break the foam down with water mist, although this method adds to the water damage.

Either fresh or salt water may be used for the manufacture of foam.

GENERAL.
Gas fires. Water should be used to cool the containers but gas flames should be extinguished only by cutting off the gas supply. Operators should take up a position where they will be protected from exploding cylinders.

Fires in the open. Try to keep to windward of the fire to avoid smoke and heat. Heat rises, so a crouching position is to be preferred to an upright one.

Fires in compartments. Except when using foam, every effort should be made to restrict the air by closing all doors, hatches, ports, windows, ventilators, and most important of all, ventilator trunks. All ventilating fans should be stopped. When entering a compartment, keep as close to the deck and bulkheads as possible, avoid standing up and the centre of the compartment.

Always have the hose with full fire pressure ready before opening the door of a compartment containing a fire and play the hose into the compartment as the door opens. If there has been a build-up of heat, the cold air entering the bottom of the doorway will cause a sheet of flame to explode from the top of the doorway, unless a strong jet of water is there to kill it as the door opens.

Fixed installations. Many ships are now fitted with fixed installations for fire fighting in compartments and engine rooms. The two main types in use are as follows: —

Carbon dioxide ($C.O._2$). A number of bottles containing carbon dioxide are located at a strategic point within the compartment. In the event of fire, the gas is released and since it is heavier than air, fills the lower part of the compartment, thus extinguishing the fire. All ventilation should of course be sealed off. Trouble arises after the fire has been extinguished because the heat is still present and if air is allowed to enter the compartment before it has cooled, the fire may well relight itself.

Sprinklers. In this method water is piped under pressure to spaced sprinkler heads in the deck-head. In the event of fire, the heat generated will burst a frangible glass bulb sealing the sprinkler head, thus allowing water to sprinkle onto whatever may be below.

Fire alarms. There will be at least two and possibly more manual fire alarms placed at strategic points on the ship. In the event of a person discovering a fire, he should immediately sound the alarm from the nearest point.

The most reliable fire alarm is YOU, if you smell smoke, see smoke coming from a ventilator or crack, or if a bulkhead is abnormally warm, investigate. If you are of the opinion that it could be a fire, sound the alarm. Do not be afraid of giving a false alarm, it is far better to give a false alarm than to risk a serious fire.

There are a number of automatic fire alarms on the market, which are suitable for ships. With some, air is drawn continuously from compartments and passed through glass tubes, in which smoke, if there is any, may be seen by an Officer on the bridge. Others consist of two strips of different metals, in the case of fire, heat expands the metals at different rates, this in turn causes an alarm to be rung. It may also be arranged to turn on a fixed installation in the compartment concerned. Again these two methods may be combined in one installation. Another type works with a sort of magic eye. Smoke in the atmosphere obscuring the magic eye will have the same effect and sound the alarm. Unfortunately these may be triggered off by dust or other impurity in the air.

Glass Bulb Type Sprinkler Heads

LIFESAVING APPLIANCES.
Fresh air breathing apparatus.

Several types of fresh air breathing apparatus have been approved for use on board ships, all of which are relatively simple to operate, even by practically untrained personnel.

The Antipoys Marine Apparatus manufactured by Messrs. Siebe Gorman, has a face mask and 60 feet (18m) of air hose attached to a lifting harness and 70 feet (21m) of hemp covered wire rope lifeline. It is suitable for breathing in smoke or toxic atmospheres on board ship over short distances. The fresh air end of the hose is secured to a belt worn by an attendant stationed in fresh air. This allows the attendant to ensure that the free end of the air hose remains in the fresh air. A system of signals can be used on the lifeline to keep wearer and attendant in communication with each other. Air is drawn into the apparatus by the wearer's own inspiratory efforts.

The Vista Smoke Apparatus also manufactured by Messrs. Siebe Gorman, has 120 feet (37m) of air hose attached to a foot bellows and 130 feet (40m) of lifeline. If the bellows cease to function the wearer is still able to draw air through the air line by his own inspiratory efforts. For use in firefighting or where heavy work is undertaken in a toxic atmosphere. Some training is needed by the attendant before he is able to gauge the correct speed at which the bellows require to be operated in order to give the operator maximum comfort when working.

The Bloman Rotary Breathing Apparatus is somewhat similar but is capable of supplying air to two operators by means of a rotary blower.

Compressed Air Breathing Apparatus.

Again several types of compressed air breathing apparatus have been approved for use aboard ships. However, it is essential that no one should attempt to wear this type of apparatus unless they have received a thorough training in its use. Neither should anyone attempt to use it without a lifeline and an attendant standing by.

A short duration compressed air breathing apparatus consists of a mask and an air bottle (which is connected to the mask) slung from the shoulder and gives complete protection against smoke and all toxic atmospheres for about ten minutes. It is ideal for use by an attendant evacuating a person who has been overcome by smoke or gas. Being easily and quickly donned and adjusted.

The Air Master Marine Compressed Air Breathing Apparatus Manufactured by Messrs. Siebe Gorman consists of a mask, harness, back plate, an air bottle and a helmet together with a lifeline 120 feet (37m) long and is supplied in a wooden or steel chest which may be mounted on a bulkhead. Two spare air bottles are supplied and kept in the chest. A 10 litre air bottle has a hard work endurance rate of about 25 minutes or 60 minutes at rest.

Both fresh air and compressed air apparatus can be fitted with a telecommunication system which can be used to keep the attendant and operator in communication with each other. The equipment consists of a full vision face mask fitted with a microphone and earpiece and a cable attached to either the lifeline or air hose of a fresh air apparatus. The attendant has a twin earpiece headset and a boom microphone connected to a control box carried on a neck sling. The control box is equipped with an on/off switch and a "press to talk" spring loaded button. The operator's microphone remains switched on throughout so that he can be heard at all times by the attendant. The system is powered by an electric battery.

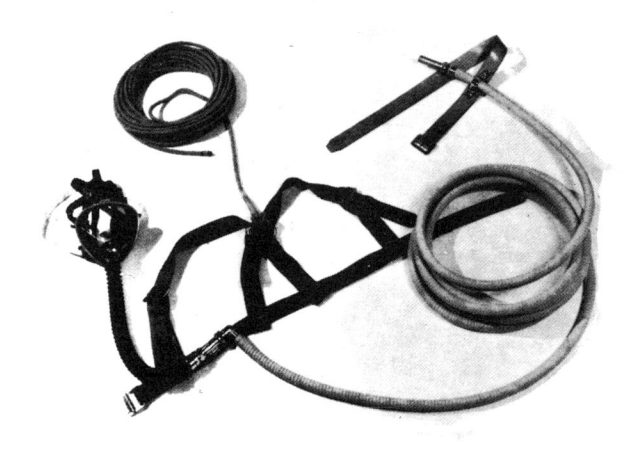

Antipoys Fresh Air Marine Apparatus

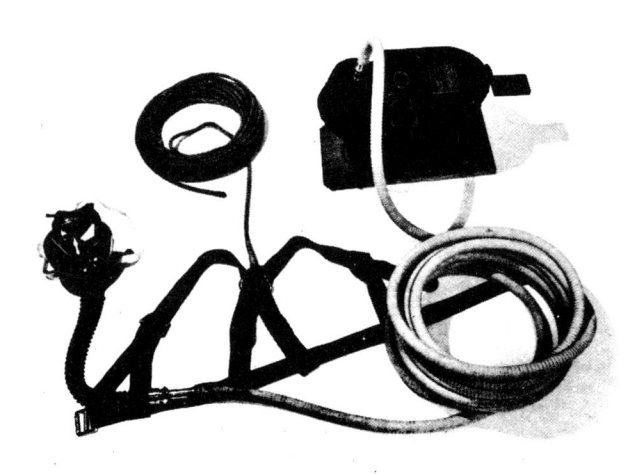

Vista smoke mask apparatus (for use on ships)

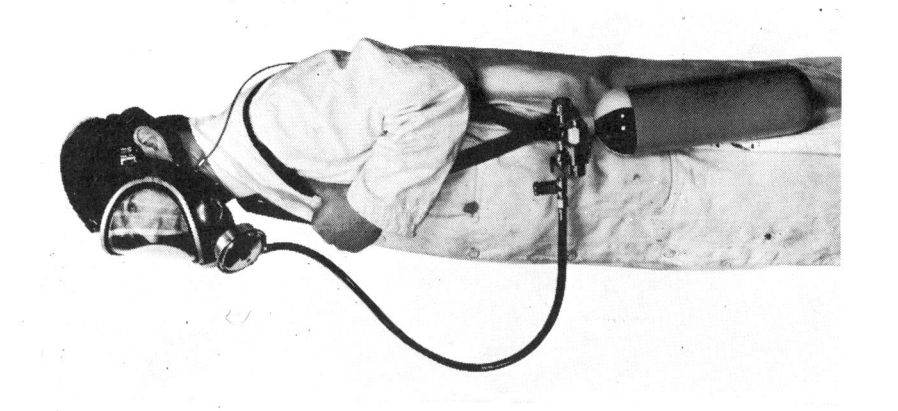

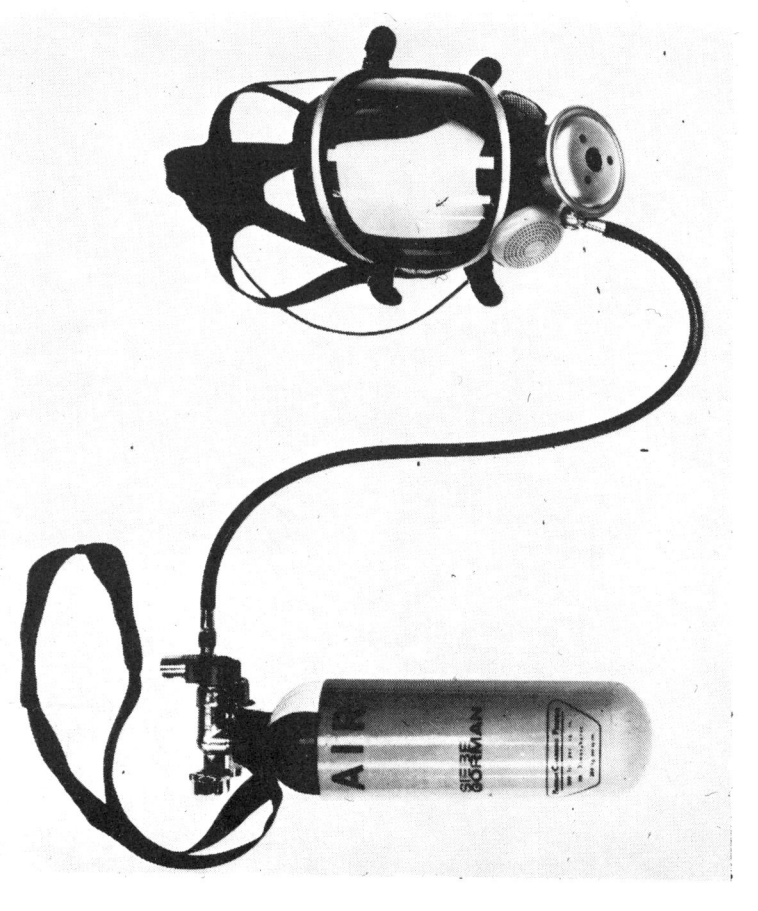

Short duration compressed air breathing apparatus

AirMaster marine compressed air breathing apparatus

Puretha gas respirator

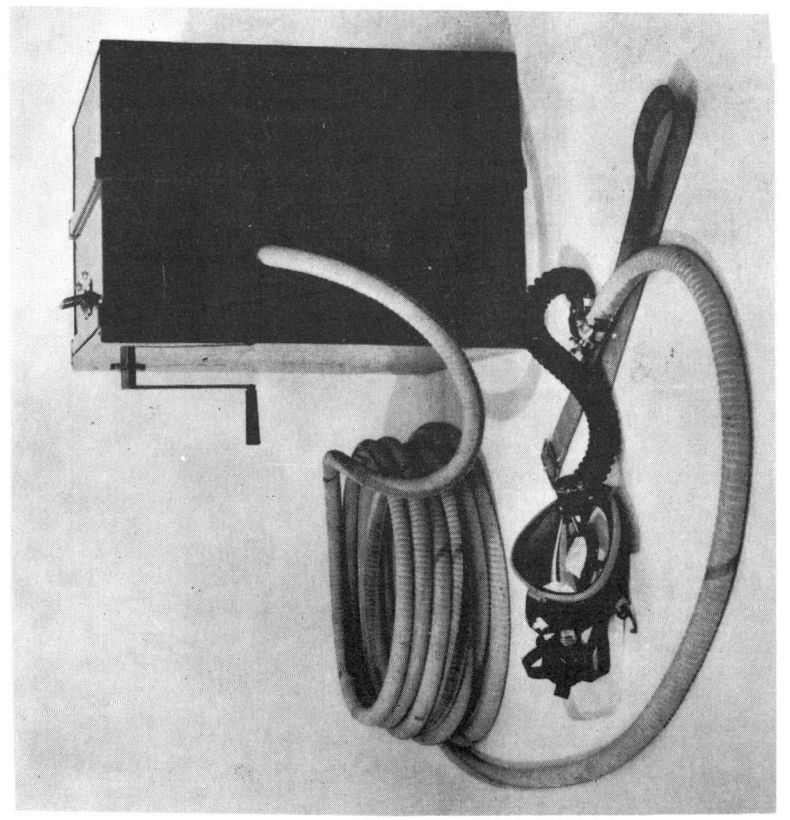

Bloman rotary blower breathing apparatus

GENERAL.

It must be appreciated that breathing apparatus is supplied not only for use in smoke when fighting a fire but also for the use of any person who may have cause to enter any compartment that might contain a toxic or inert gas or which has been unventilated for some length of time and in which the atmosphere may be suspect.

Owing to the possibility of toxic air entering a face mask, it is advisable that breathing apparatus should not be worn by a person who has a beard.

When wearing breathing apparatus, if gas or smoke is smelt inside the mask, leave the area immediately and ascertain the cause. e.g., torn or damaged rubber, cracked visor, etc. In the event of a visor cracking when wearing compressed air breathing apparatus, open the by-pass valve immediately and leave the contaminated area at once.

Gas Masks.

Gas masks are provided for the use of crews on liquid gas carriers and other ships carrying hazardous cargoes. A gas mask is of no use whatsoever in an atmosphere devoid of oxygen. It can however, depending on the filters incorporated in the mask, filter toxic gases from the atmosphere and so, provided the atmosphere contains sufficient oxygen, render the air safe to breathe. In view of this, gas masks should not be used in tanks, pumprooms and other confined spaces, neither are they suitable breathing apparatus for wear in a smoke laden atmosphere. The filters in gas masks should always be renewed immediately after use, as their life is limited. Care must also be taken to ensure that the filters in the mask are in fact capable of absorbing the toxic substances in which the operator is working.

Dust Masks.

For use when working in dust, in a safe atmosphere. Filters must be renewed after use.

PROTECTIVE CLOTHING.

The protective clothing carried aboard a ship will depend upon both the trades she follows and the cargoes she carries. Vessels carrying hazardous cargoes will carry full protection suits. Vessels carrying general cargo will carry little more than a number of protective helmets, goggles, safety belts and perhaps some rubber gloves. It is therefore up to every seaman to provide himself with a pair of suitable working gloves and a pair of suitable protective boots or shoes.

A gas protection suit designed for work in highly toxic conditions such as exist when loading or discharging a cargo of ammonia or chlorine is manufactured by Messrs. Drager Normalair Ltd. Made of a light but allway stretch material, it is of one-piece construction with entry effected by means of a special frontal zip fastener. The trousers being adjustable by means of special braces, it has a face mask bonded to the material containing a pneumatic seal and speech diaphragm allowing easy fitting of an air-line or self contained breathing apparatus and a telecommunication set. It may be provided with gas tight cuffs or have boots and gloves. When boots and gloves are worn separately it is important to tuck the gloves inside the sleeves and the boots inside the legs, to prevent liquid running down onto hands and feet.

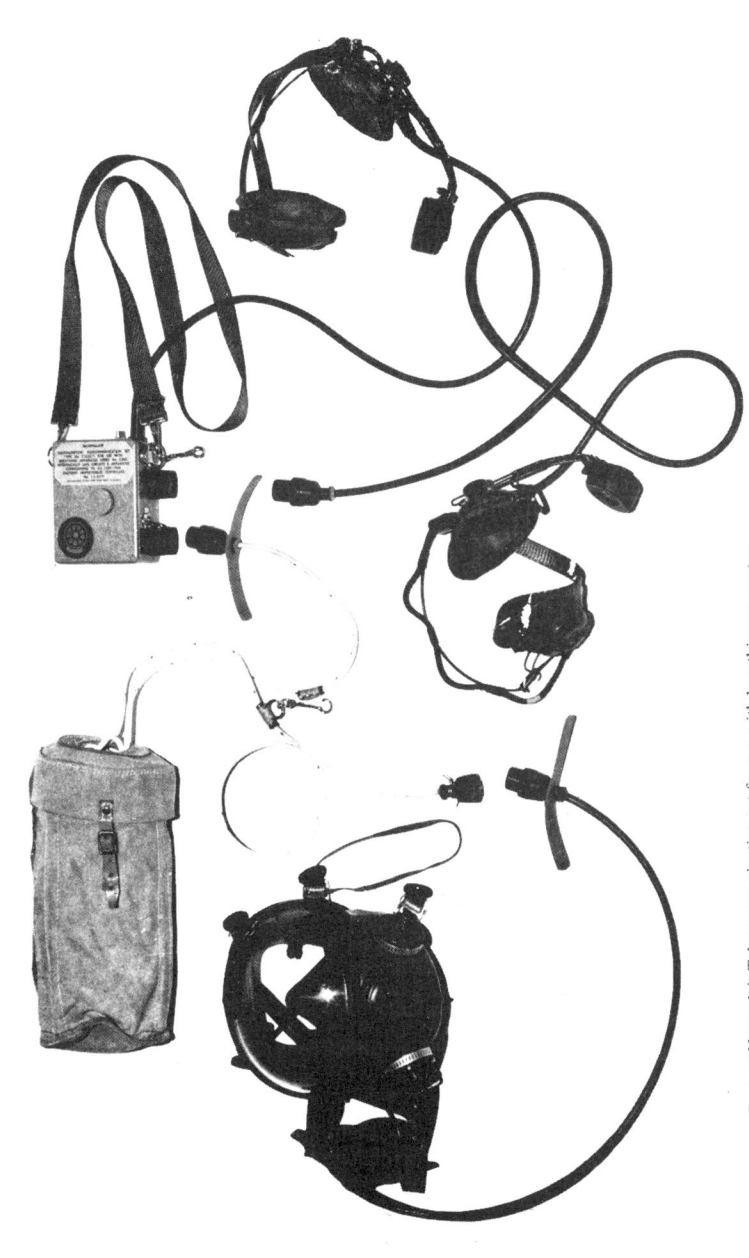

Draeger Normalair Telecommunication set for use with breathing apparatus

Draeger Normalair Gas Protection Suit showing method of dressing and zip fastener.

Draeger Normalair Gas protection suit being worn.

For less hazardous cargoes a protective boiler suit with elasticated legs and cuffs and a hood may be supplied together with a gas mask which is kept in the ready position when the suit is worn. The protective boots should in all cases be non-static.

Protective helmets should always be worn whenever working in a ship's hold, when opening or closing hatches, raising or lowering derricks and whenever cargo is being worked or work is being carried out aloft or below.

Goggles should be worn when painting with a spray gun, chipping or scraping, using a grindstone, electric or pneumatic tools, working in the vicinity of acid, dust, corrosive or welding processes and when letting go the anchor.

Loose gloves should be worn when handling wire ropes, rough castings, wood and under arctic conditions. When handling oils and chemicals suitably protective gloves should be worn.

Footwear. All footwear should have a strong non-skid sole and cover the whole foot. Footwear containing a protective in-sole plate and toe-cap, should always be worn under the same conditions as protective helmets. Open-toed footwear is most unsuitable for any shipboard use and should never be worn. On ships carrying hazardous cargoes anti-static footwear should be worn.

Orders referring to protective clothing are not issued for fun, no matter how unnecessary they may seem. Under no circumstances should any member of the ship's crew attempt to evade the wearing of protective clothing when carrying out duties whereby the regulations specify that protective clothing is to be worn, or at any other time when it may be prudent.

Gas tight torches. All ships carrying cargoes which are liable to give off inflammable vapours will be supplied with these torches which are battery operated. Owing to the dangers occasioned by sparking, no other torch or light is to be used when there is any danger from an inflammable or explosive gas in the vicinity.

Safety belts. are supplied by the ship and are always to be worn when working aloft, most especially when overhauling gear aloft, when working overside and when rigging an accommodation ladder or shifting boards.

Instructions for the use of the "Protector" Nylon Satety Harness Belt.
The belt has been designed and tested for safety and comfort and is manufactured from high tensile nylon to B.S.S. 1397. Look after it for your own good.
For correct fitting, the arms go through the shoulder straps to keep the belt well up on the chest. Pass the flap end of the belt up and around the milled slider of the buckle, draw reasonably tight and take the free end through the loop provided to keep it in place. Check that web locks in the buckle.
It is important never to have more loose rope than is needed and the maximum direct fall must never exceed four feet.
The Nylon Suspension Rope has a high degree of elasticity, and the back pocket of the belt contains rope which passes through the Shock Absorber Friction Buckle in the event of a fall. Do not withdraw any of this rope or the full effectiveness of the shock absorbing device will be lost. With the maximum free

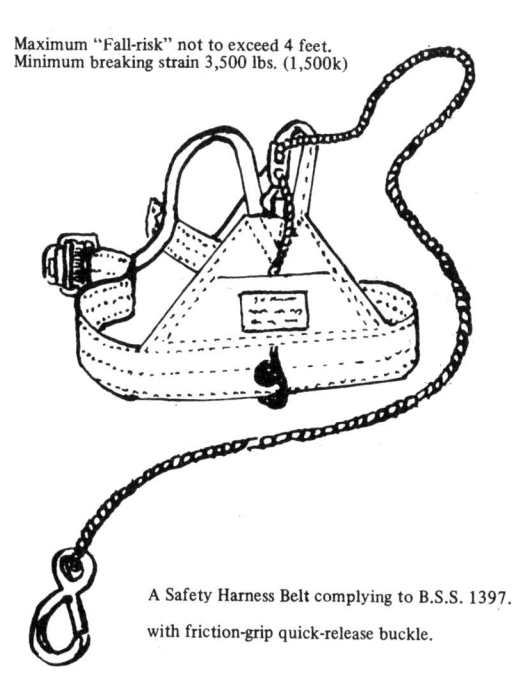

Maximum "Fall-risk" not to exceed 4 feet.
Minimum breaking strain 3,500 lbs. (1,500k)

A Safety Harness Belt complying to B.S.S. 1397.

with friction-grip quick-release buckle.

The "NEIL-ROBERTSON" stretcher.

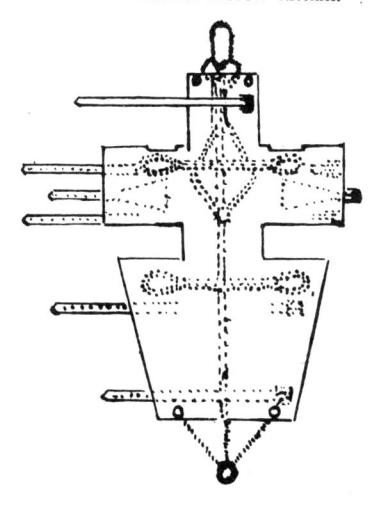

Bamboo strips (not shown)

fall of 4 feet (1.2m.), plus the shock absorbing device rope, together with the high elasticity of the nylon rope itself, the effective distance of final arrest is approximately 8 feet (2.4m.). Care should be taken that when there are obstructions at less than this distance and the suspension rope kept as short as possible.

Store the belt away from sharp tools, chemicals, etc., and if it gets wet, dry it naturally by hanging up in an airy room. Never use artificial heat.

Always examine the belt and lanyard before and after use for signs of cuts or chafing. Have it replaced if defective in any way.

ATMOSPHERE SAMPLING EQUIPMENT.

All ships carrying hazardous cargoes now carry air sampling equipment, provided for the purpose of sampling air in compartments. It must not be presumed that simply because an air sample has been shown to be satisfactory, that a compartment is in fact absolutely safe. Toxic and explosive gases and vapours, many of them being heavier than air, can and do remain in unsuspected pockets, even after a compartment has been well ventilated. Never go into a suspect compartment unless you have someone with breathing apparatus standing by.

Fixed installations. Some ships may be equipped with a number of fixed explosion proof detector filament heads fitted in suitable positions within compartments. The detectors are connected to a remote metering system situated in a comparatively safe position, from where the condition of the atmosphere in any of the fitted compartments may be sampled for any gas or vapour the detector equipment is calibrated to detect.

Portable equipment, of which there are a number of varying types, is at the time of going to press, carried by the majority of ships employed in the carriage of hazardous cargoes.

Davy-lamp. An oil burning safety lamp carried on some ships (particularly coal burning ships) and used to determine that the atmosphere in a compartment (especially a coal bunker or water ballast tank) is safe to breathe. A lack of oxygen will dowse the light.

Explosimeter. The Explosimeter 2E manufactured by the Mine Safety Appliances Company Limited, is electric battery operated and is carried to the compartment it is desired to test in a small case, with a sampling tube attached. The sampling tube is inserted into the compartment and the atmosphere from the compartment is drawn into the explosimeter by means of an aspirator bulb.

The instrument will indicate what percentage of the lower explosive limit (L.E.L.) of combustibles are present. Where concentrations are likely to be in excess of the L.E.L the Mine Safety Appliances "Gascope" can be used, as the 2E Explosimeter, and the percentage of gas present in the atmosphere read directly. It is however necessary to know the gas present so that a correctly calibrated "Gascope" can be used. Neither instrument is suitable for the detection of inert gases, or measuring the percentage of oxygen present in the atmosphere.

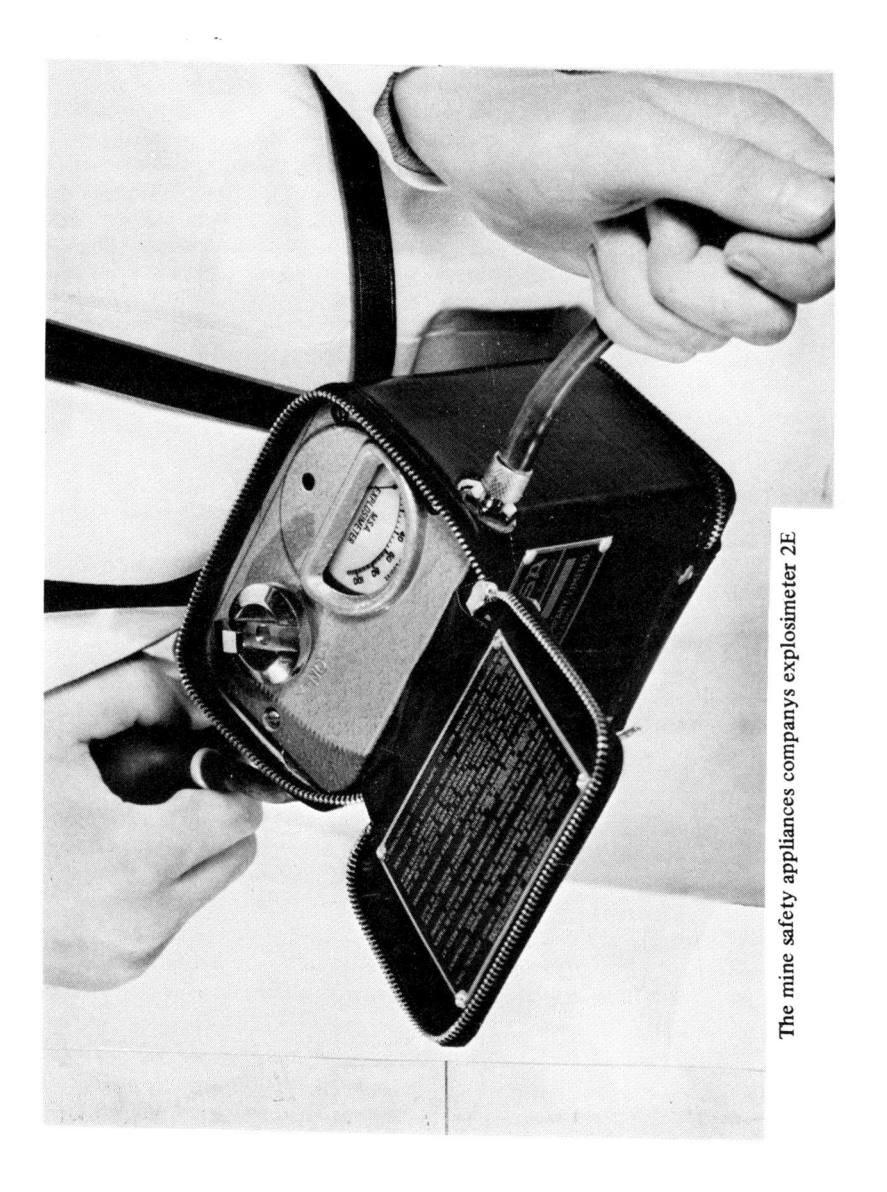

The mine safety appliances companys explosimeter 2E

FYRITE Gas Analyser

Draeger Normalair Gasalarm suitable for the detection of leaks of all combustible gases or for testing atmosphere up to the lower explosive limit. Can also be calibrated for n-Pethane or Methane

Fryrite Gas Analyser manufactured by Messrs Shandon Southern, which uses a chemical fluid, can measure the amount of carbon dioxide ($C.O._2$) or alternatively (depending upon the fluid contained in the analyser). The amount of oxygen (O_2) present in the atmosphere of a compartment, when air from the compartment is drawn, via a sampling tube and aspirator bulb, into the analyser.

Multi Gas Detector manufactured by Messrs Draeger Normalair Ltd., uses a bellows, sampling tube and glass tubes. Air drawn from a compartment by means of a bellows and sampling tube, passes through a glass tube. The amount the atmosphere is contaminated can then be determined by discolouration of a chemical contained in the glass tube, much like a breathalyser. However, each glass tube is only capable of detecting a particular gas, so that differently filled tubes have to be used to detect different gases, but most, if not all the toxic and explosive gases and vapours that are liable to be found in a compartment aboard ship, can be detected by the Multi Gas Detector, provided the correct tubes are used to detect the gas present.

THE "NEIL-ROBERTSON" STRETCHER.

A "Neil-Robertson" stretcher is required to be carried on all ships. Constructed of stout canvas into which strips of bamboo have been inserted lengthwise, it is designed for lifting casualties in any position through small hatches, man-holes, ventilators, up and down companionways, round corners and lifting or lowering from heights.

The casualty is strapped tightly into the stretcher. A ring at the top is used for hoisting. A lined strap is provided at the top for passing round the casualty's head if he is unconscious. A length of rope may be attached to the ring at the foot of the stretcher for guiding it. The casualty's feet are placed on the rope attachments at the base—these serve as stirrups. The four loops at the side are for four people to carry the casualty.

RESUSCITATORS.

The Respirex Ox-Vital, Oxygen Mask, manufactuted by Respirex Ltd., is now carried on a large number of ships, including tankers and chemical tankers. It is for use when the casualty's breathing is impaired or has ceased, particularly in cases of:— drowning and accidents in water, shock from electrical and accident causes, gassing and poisoning, heart attack and circulatory collapse, respiratory troubles, etc., and can be used by anyone, no particular training being necessary. It is probably better and certainly easier than the "kiss of life" which is not really capable of being carried out satisfactorily by untrained personnel.

Briefly, the Ox-Vital mask consists of a mask and a bellows containing a coiled tube in which sufficient oxygen to last for about 15 to 20 minutes is stored.

INSTRUCTIONS FOR USE are quite simple
1. Release the clips and remove the lid.
2. Unscrew the base and pull ring inside oxygen store to release oxygen.
3. Replace base by pressing in firmly and turning.
4. Now medical oxygen will flow continuously into the bellows for about 20 minutes. The rate of flow of the oxygen will reduce slowly as time progresses.
5. The strap under the base of the oxygen mask should be used to give a better grip for the hand.

6. Even after the oxygen has ceased to flow, resuscitation procedure can be continued, as the bellows will pump air into the lungs under these circumstances.
7. If the patient is conscious and able to sit up, place the soft plastic face pad with the narrow part uppermost about ½ inch (13mm) away from the face of the patient. The patient should breathe in and out deeply and evenly.
8. If the patient is unable to sit up, conscious or unconscious, but the NATURAL BREATHING APPEARS TO BE ADEQUATE, lay the patient in the most comfortable position. Place the soft plastic face pad with the narrow part uppermost, about ½ inch away from the face of the patient and allow him to breathe naturally.
9. IF THERE IS A TOTAL STOPPAGE OF BREATHING OR BREATHING SEEMS TO BE VERY WEAK. FOLLOW THESE INSTRUCTIONS WITH CARE.
   a. Lay the patient down on his back, put a cushion or similar object **under the shoulder blades** (never under the head or neck). In this position the breathing passages should be open and free of natural obstructions.
   b. Check that this is so by looking into the mouth, and if necessary, move the tongue forward away from the throat and remove any foreign matter from the mouth and throat. Remove false teeth and spectacles.
   c. Place the soft plastic pad with the narrow part on the bridge of the nose, and by light but firm downward pressure, force the oxygen over the mouth and nose into the lungs.
   d. Allow the bellows to expand and lift away from the face. Check that the tongue is forward and that there is no foreign body in the mouth and throat.
   e. Replace the mask over mouth and nose and repeat. Continue this procedure evenly, taking 3 to 4 seconds over each cycle.
   Look at the patient's chest. It should go up and down with breathing in and out. If it does not, or you feel any resistance, lift the mask away from the face and check again that the breathing passages are free of obstruction. You may help with the artificial respiration by slightly pressing your hand on the chest after breathing in and during the breathing out phase. Continue until the patient can breathe by himself.
IN CASES OF DROWNING etc., where a continuation of resuscitation may be successful, continue the above procedure even after the oxygen has ceased. The bellows will pump air into the lungs under these circumstances.
10. To fit replacement oxygen cartridge, unscrew the base, pull out the empty cartridge and replace with a full one. Replace base. Retain empty cartridge for refilling.

VARIOUS TERMS APPLICABLE TO SAFETY PRECAUTIONS.

Asphyxia.	Death due to lack of oxygen—suffocation.
Atmosphere.	The air we breathe.
Casualty.	A person who has met with an accident.
Combustible.	Will burn easily and rapidly.
Corrosive.	To eat away the flesh or a metal by chemical action.
Flash point.	The lowest temperature at which a gas or vapour will catch fire when in contact with a spark or flame.

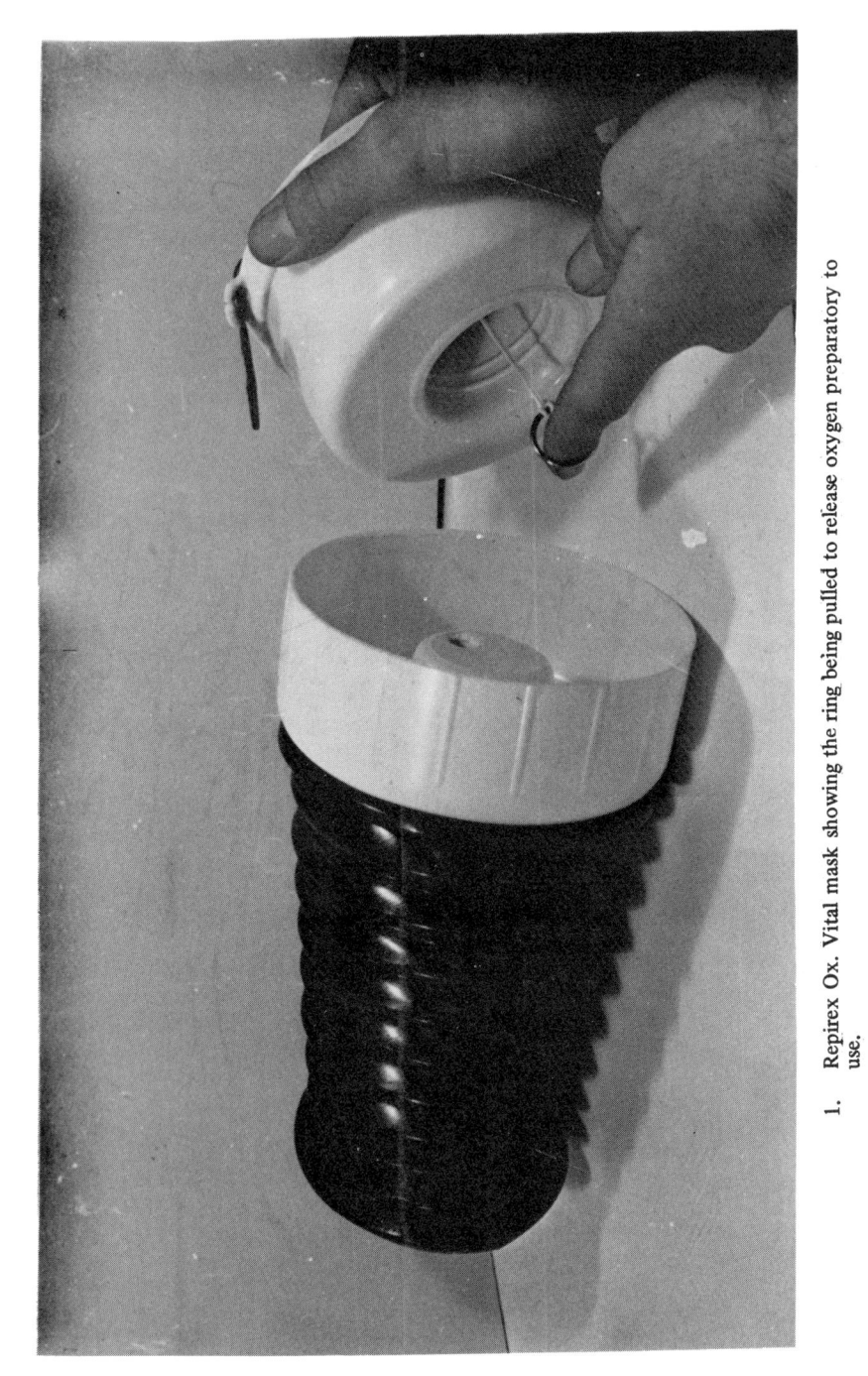

1. Repirex Ox. Vital mask showing the ring being pulled to release oxygen preparatory to use.

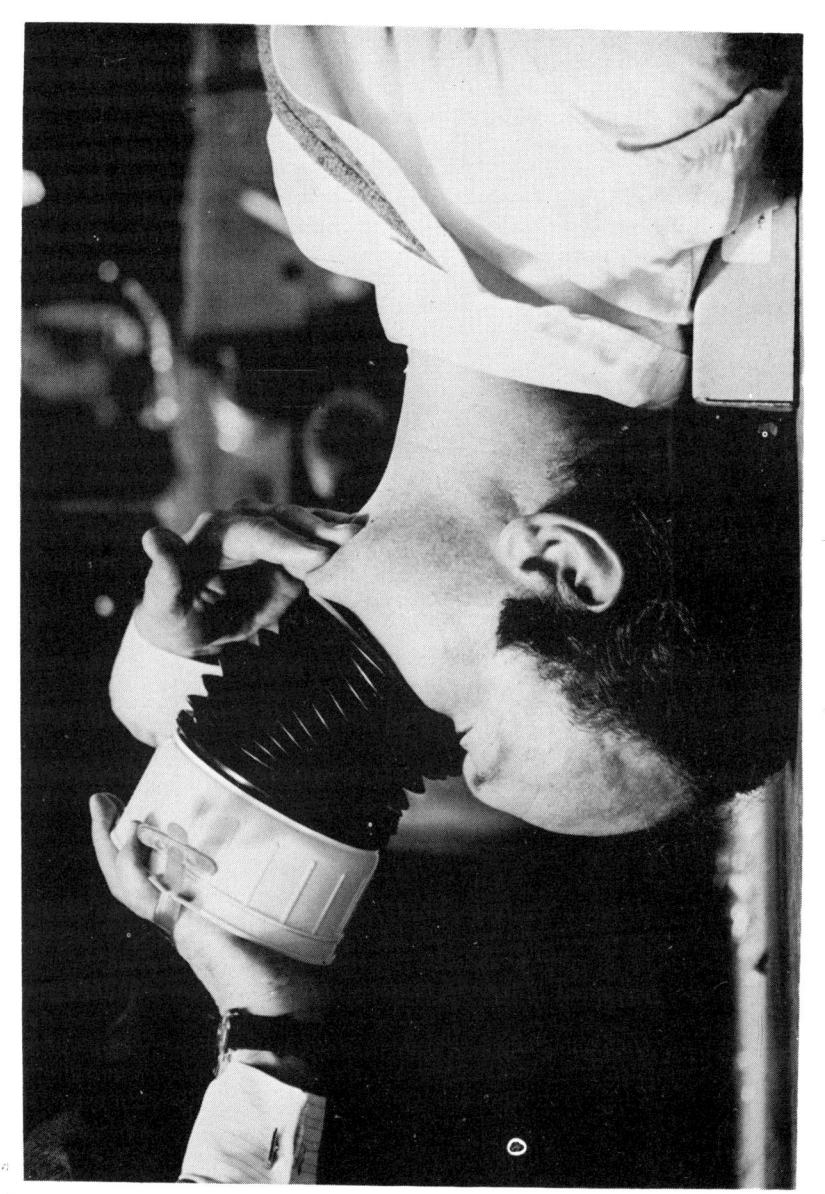

1A. Mask in position and making the casualty breathe in.

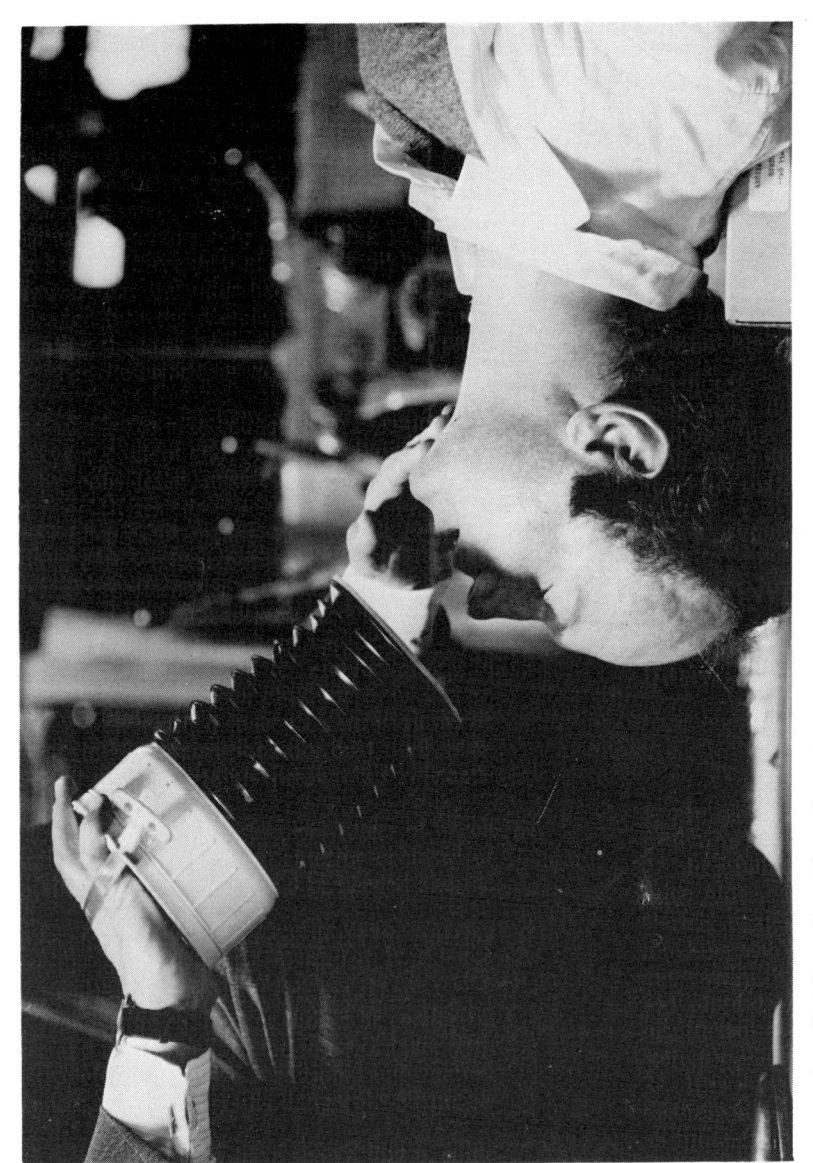

2. Position of mask to allow the casualty to breather out, and his condition to be observed.

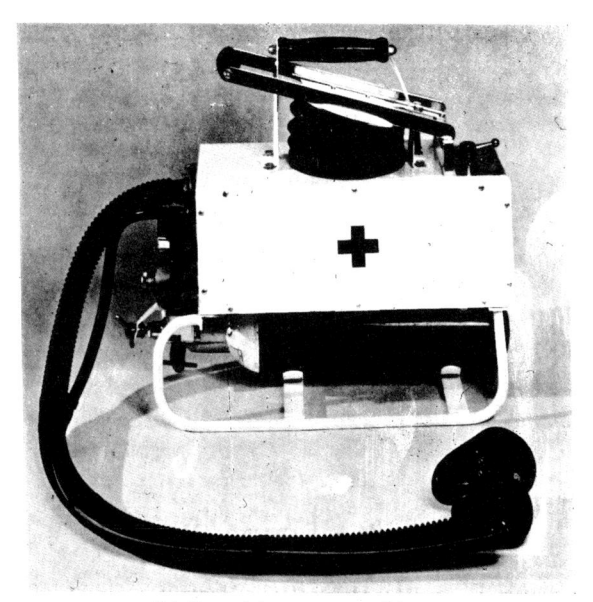

Siebe Gorman Automan oxygen resuscitator

Draeger Normalair Oxyalarm. When worn by an operator in a suspect atmosphere will give an automatic audible alarm should the oxygen content become either deficient or enriched.

DANGEROUS GOODS LABELS (IMCO)

Label 1A	EXPLOSIVE — 1
Label 1B	FIREWORKS (SHOP GOODS) — 1
Label 2A	INFLAMMABLE GAS — 2
Label 2B	POISON GAS — 2
Label 2C	COMPRESSED GAS — 2
Label 3	INFLAMMABLE LIQUID (FLASH POINT) — 3
Label 4A	INFLAMMABLE SOLID — 4
Label 4B	SPONTANEOUSLY COMBUSTIBLE — 4

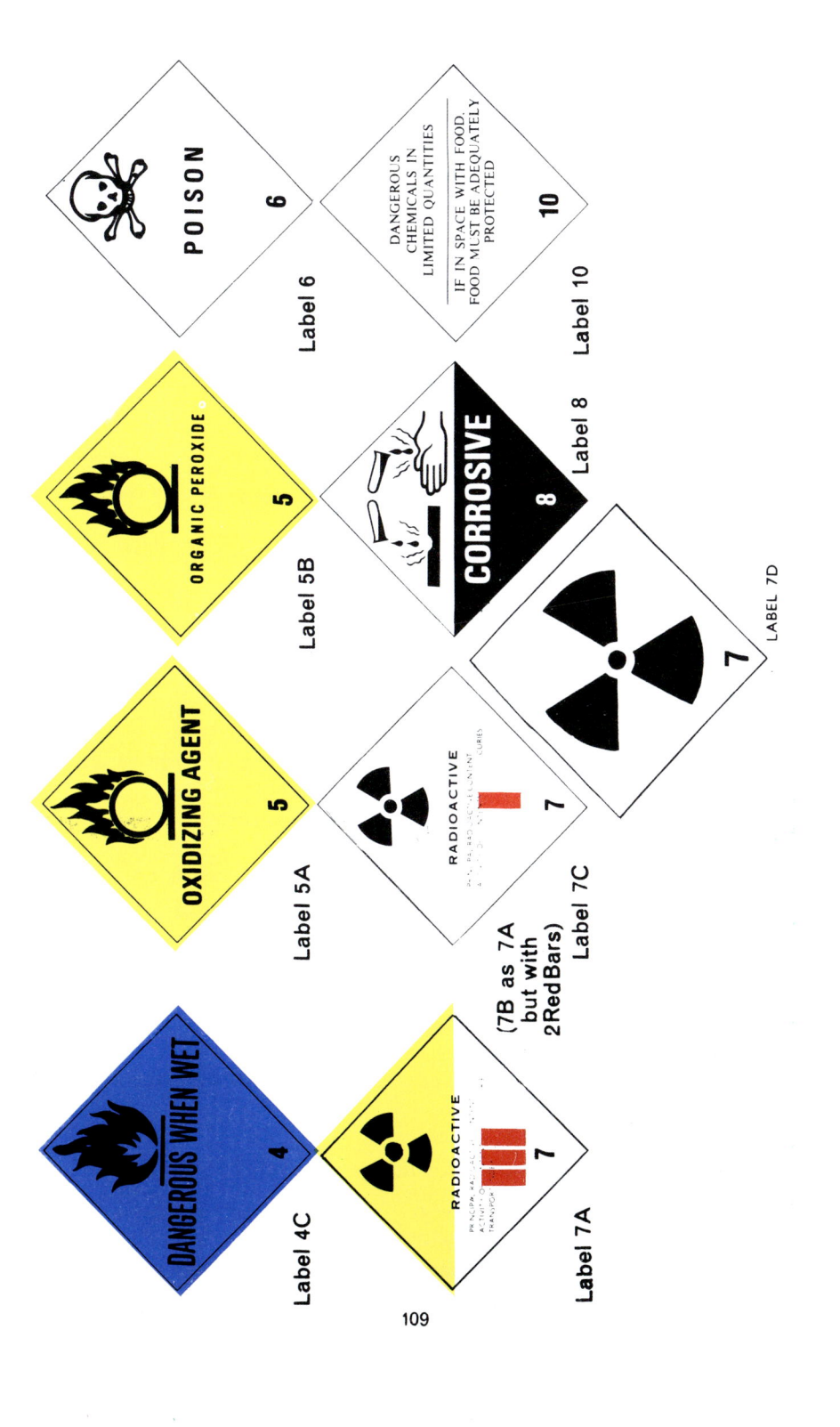

POISON 6	DANGEROUS CHEMICALS IN LIMITED QUANTITIES / IF IN SPACE WITH FOOD, FOOD MUST BE ADEQUATELY PROTECTED 10		
Label 6	Label 10		

ORGANIC PEROXIDE 5 — Label 5B

CORROSIVE 8 — Label 8

7 — LABEL 7D

OXIDIZING AGENT 5 — Label 5A

RADIOACTIVE 7 — Label 7C

(7B as 7A but with 2 Red Bars)

DANGEROUS WHEN WET 4 — Label 4C

RADIOACTIVE 7 — Label 7A

109

Hazardous.	Dangerous.
Ignition point.	The lowest temperature at which a gas or vapour will catch fire spontaneously (without being in contact with a spark or flame).
Inert gas.	A non-toxic gas that will not sustain life or fire.
Inflammable.	Will catch fire easily.
Oxygen.	A gas in the atmosphere which is necessary to sustain both life and fire.
Radio-active.	Gives off harmful rays which can damage any person with whom they come in contact.
Resuscitation.	Bringing back to life when nearly dead.
Respiration.	The act of breathing.
Spontaneous combustion.	Catches fire without coming in contact with fire.
Static.	An un-earthed charge of electricity.
Toxic.	Poisonous.

ACCIDENT PREVENTION.

Never use a fire bucket, fire hose or fire sand for any purpose other than firefighting.

Never, when in the vicinity of appliances using cylinder gases (L.P.G.) such as propane or butane, check for leaks with a naked flame.

Never leave a fire extinguisher or fire bucket empty or out of place. Re-charge and return it to its position as soon after use as possible.

Never throw a lighted match or cigarette end away. Put in in a proper receptacle.

Never smoke in a no-smoking area.

Never smoke in bed. (Have your last cigarette before going to bed).

Never enter a peak, tank or other compartment where the air might be suspect without well ventilating first.

Never enter any suspect compartment without having an attendant and breathing apparatus at the entrance.

Never leave the light on in an empty cabin or compartment. Switch it off.

Never leave electrical equipment without first disconnecting it or switching off at the MAINS.

Never cover an electric light bulb closely with material put there to act as a shade. Leave ample room for air circulation.

Never lay or hang clothing or anything else where it can fall onto an electric fire, radiator or electric light bulb.

Never use old flex or a multi-point adaptor to bring electrical equipment into use.

Never use defective electrical equipment.

Never interfere with the ship's electrical fittings.

Never spray an aerosol near a naked flame or electric fire.

Never place an aerosol where it can become overheated.

Never carry strike-anywhere or wax-impregnated matches on a ship.

Never carry a lighter on a tanker or liquid gas carrier.

Never wear defective protective clothing.

Never enter a smoke filled compartment without breathing apparatus and an attendant standing-by.

Always keep the doors, ports and ventilators of unoccupied cabins and storerooms shut.

When an attendant is standing-by a person who has entered a pump room, tank or other suspect compartment, the attendant should keep a close watch on that person. Should the person appear to act drunkenly, it is probably a sign that he is unknowingly being overcome by toxic fumes and he should therefore be called out immediately and before he collapses. If the person in the suspect compartment should collapse, it is the duty of the attendant to summon aid as quickly as possible by sounding the appropriate alarm. On no account should an attendant ever enter a suspect compartment leaving himself unattended, either with or without breathing apparatus.

When a person is overcome by toxic fumes, it is of vital importance to remove him from the suspect compartment as quickly as possible. As few as four minutes may mean the difference between life and death. First aid, together with any other medical attention is given after the patient has been removed from the toxic atmosphere.

Merchant Shipping Notice No. M.946

FIRES INVOLVING ELECTRIC HEATING OR DRYING EQUIPMENT

Notice to Owners, Masters, Officers and Crew Members of Merchant Ships, Owners, Skippers and Crew Members of Fishing Vessels, Shipbuilders and to the Builders of Fishing Vessels

This Notice supersedes Notice M.570

1. In recent years there have been a number of fires involving electric heaters or drying equipment incorporating an electric heater.
2. In a number of cases fires have resulted from items of clothing, bedding or other objects being placed too close to, or inadvertently falling onto, unguarded electric heaters. In another incident a drying cabinet was overfilled thereby blocking the ventilation apertures with the result that the contents overheated and caught fire. Other incidents have involved portable electric heaters installed as temporary heating during very cold weather or for use in cold climates when insufficient attention was given to the positioning of the heaters so that they were too close, or immediately below, flammable objects.
3. It is important that all fixed electric heaters are fitted with suitable guards securely attached to the heater and that the guards are maintained in position at all times. Temporary arrangements to hang clothing above the heater or to dry clothing on the heaters should not be permitted and drying of clothing should only be carried out by using suitably designed equipment.
4. When using drying cabinets or similar appliances care should be taken so that the ventilation apertures are not obscured by overfilling of the drying space.

As the ventilation apertures of drying appliances may become blocked due to accumulations of fluff from clothing any screens or fine mesh covers associated with the ventilation apertures should be regularly inspected and cleaned.

5. The use of portable heaters should be avoided. However, if they are used with the ship in port as temporary heating during repairs and as additional heating during inclement weather, the heaters should not be positioned on wooden floors or bulkheads, carpets, or linoleum without the provision of a protective sheet of a non-combustible material. Portable heaters should be provided with suitable guards and care should be exercised when positioning the heater in relation to furniture and other fittings in the cabin or other space. Again, drying arrangements in relation to these heaters should not be permitted.

6. The construction and installation of electric heaters in merchant ships and fishing vessels should take due account, as appropriate, of the requirements of regulation 47(6) of the Merchant Shipping (Passenger Ship Construction) Regulations 1980, Regulation 10(11) of the Merchant Shipping (Cargo Ship Construction and Survey) Regulations 1980, Regulation 14(5) and Schedule 6, paragraph 7(3) of the Merchant Shipping (Crew Accommodation) Regulations 1978, Regulation 14(5) of the Merchant Shipping (Crew Accommodation) (Fishing Vessels) Regulations 1975 as supplemented by paragraph 2.10.1 of the Survey of crew accommodation in merchant ships—Instructions for the guidance of surveyors and paragraph 2.10 of the Survey of crew accommodation in fishing vessels—Instructions for the guidance of surveyors.

7. Permanent electric heaters are normally supplied with installation instructions by the manufacturers and these should be carefully complied with.

8. Attention is also drawn to chapter 2 and chapter 26 of the Code of safe working practices for merchant seamen.

Merchant Shipping Notice No. M.908

HEATING APPLIANCES BURNING SOLID FUEL

Notice to Owners, Masters Officers and Seamen of Merchant Ships and Yachts; and to Owners, Skippers and Crews of Fishing Vessels

This Notice supersedes Notice M.575

1. A number of deaths have occurred where seamen have been asphyxiated by fumes given off from heating appliances burning solid fuel.

2. Where any such oxygen-consuming appliances, whether permanent or temporary, are in use on board the following precautions should be strictly observed:
    (i) **Heating appliances.** The heating appliance itself should be regularly and frequently examined to ensure that it is maintained in a condition which will enable it to function properly. Particular attention should be paid to the following points:
        (*a*) doors where fitted should be capable of beng completely closed and where mica panels are incorporated these should be intact;
        (*b*) draught regulators should function properly;
        (*c*) the separate components forming the appliance must fit pronerly and be free from cracks;

(d) the flue pipe must be unobstructed by soot deposits—any horizontal (or near horizonal) lengths of flue should receive frequent attention.

(ii) **Ventilation.** In cases where deaths have occurred in the circumstances referred to above, the ventilation system has been found inefficient due to having been interfered with or neglected. It is not unusual to find inlet ventilators deliberately blocked, and butterfly and sliding vents in cabin doors have been found to be in the closed position and immoveable.

It cannot be too strongly impressed upon all concerned that adequate ventilation of accommodation, and in particular of sleeping rooms, is of primary importance and on no account must a ventilation system be interfered with so as to prevent its proper functioning.

3. It is recommended that a copy of this notice should be posted up in all spaces heated by solid fuel appliances.

Merchant Shipping Notice No. M.910

LOSS OF LIFE IN CARGO TANKS, CARGO HOLDS AND OTHER CLOSED SPACES

Notice to Owners, Masters, Officers and Crew Members of all Merchant Ships and to Shipbuilders

1. Within the last few years a number of seamen have lost their lives in cargo tanks, cargo holds and other enclosed spaces.

2. In one incident seven men were killed when they were overcome in the cargo tank of a product tanker. Initially a crew member entered the tank, which contained a few feet of slops comprising tallow, vegetable oils and seawater, prior to the commencement of a tank washing operation. This was contrary to normal procedures because the tank had not been ventilated nor had the atmosphere been tested to ascertain whether it was safe to enter. When it was realised that the crew member was in some difficulty and before it was appreciated that this was because the atmosphere in the tank was unsafe, six other crew members had rushed into the tank to assist instead of following the established and practised emergency procedures for a tank rescue. It was later concluded that the atmosphere was not only deficient in oxygen but contained quantities of other gases generated by the residue. Despite a determined rescue attempt by the remaining crew members wearing breathing apparatus, all seven men lost their lives.

3. Two other incidents occurred on bulk carriers. In the first incident a crew member unsuspectingly entered a unventilated cargo hold that contained various quantities of timber, steel and general caro loaded some two months earlier. He was seen to collapse and the three men who rushed in to help also got into difficulties because the atmosphere was deficient in oxygen. A rescue attempt was mounted by crew members wearing breathing apparatus but only one of the casualties survived. In the second incident a shore official and four crew members entered a hold containing pig iron loaded three weeks earlier. The ventilators for the hold had been kept closed during the voyage in

anticipation of rough weather and the hold had not been ventilated prior to entry. The men were well down into the hold when they experienced difficulties due to the oxygen deficient atmosphere. Although every effort to rescue them was made by crew members using breathing apparatus only one of the five who entered the hold survived. Fortunately no additional lives were lost in this case due to reckless entry by unprotected personnel.

4. In another incident four senior crew members on board a liquified gas carrier entered one of the double bottom tanks from the duct keel. While they were in the tank one of the men collapsed. Two men remained to help him while the other went to summon assistance. Because no one had been instructed to stand by the tank entrance and because no safety equipment was kept ready for immediate use in the event of a situation of this type developing, it proved difficult to mount a rescue attempt and the three men in the tank all lost their lives. Subsequent investigation showed that the tank had not been properly ventilated and that inert gas had leaked into the double bottom tank from a known defect in the bulkhead separating the tank from an adjacent void space.

5. All the above incidents have been described in detail to illustrate the following points.
   - (a) The atmosphere in any enclosed space may be incapable of supporting life. It may be lacking in oxygen content and/or contain flammable or toxic gases and should be considered unsafe unless it has been thoroughly ventilated and properly tested.
   - (b) An unsafe atmosphere may be present in any enclosed or confined space including cargo holds, cargo tanks, pump rooms, fuel tanks, ballast tanks, fresh water tanks, cofferdams and duct keels. Furthermore, it should never be assumed that precautions need not be taken for holds or tanks containing apparently innocuous cargoes.
   - (c) An enclosed or confined space should not be entered unless a "permit-to-work" or similar authority has been obtained.
   - (d) Any one who enters an enclosed or confined space to attempt to rescue a person without first taking suitable precautions not only unnecessarily risks his own life but will almost certainly prevent his colleague being brought out alive.

6. The "Code of safe working practices for merchant seamen" copies of which should be on board all ships, contains detailed advice on entering enclosed or confined spaces. The warnings contained in the Code should be heeded and the recommended procedures followed.

7. It is essential that there should be clearly laid down procedures for entering enclosed or confined spaces—preferably in the form of a "permit-to-work"—to ensure that all the necessary safety measures and precautions are taken.

8. The Code recommends that oxygen testing equipment should be carried on board all ships. Its use, maintenance and regular calibration in accordance with the manufacturer's instructions is strongly emphasised. Additional information on entry into tanks and enclosed spaces on ships carrying dangerous chemicals in bulk is contained in Notice No. M.576.

9. Investigations into incidents involving loss of life have shown that in some of the cases there were no established rescue procedures for dealing with accidents in enclosed or confined spaces. Procedures for dealing with such incidents should be formulated where they do not already exist. Regular drills should be conducted to simulate the rescue of a crew member or life sized

dummy from an enclosed or confined space—the space having been proven safe for the exercise. These drills would bring all the emergency procdures into use including, where appropriate, the use of breathing apparatus and lifelines by the rescuers, the control of their air supplies and the provision of replacement air supplies, the lowering of a resuscitator and a stretcher, the rigging of portable hoisting equipment over suitable openings and the recovery of the "unconscious person" to prove and practice the formulated emergency procedure.

10. When breathing apparatus is being used difficulty is sometimes experienced in gaining entry into enclosed or confined spaces due to the restricted size of the openings. Similar difficulties are experienced in recovering an injured or unconscious person from such spaces, or from a space with restricted access leading to another enclosed space. It is recommended therefore that with new tonnage, access hatches to or manholes or openings in enclosed or confined spaces should be of a suitable size to permit entry by a person wearing breathing apparatus and to allow the recovery of an injured or unconscious person. Furthermore the access hatch, manhole or other opening should be positioned, whenever possible, to enable an unobstructed recovery of the injured or unconscious person from the lowest part of the space.

Merchant Shipping Notice No. M.1254

USE OF SOLID CARBON DIOXIDE

Notice to Shipowners and Masters, Officers and Crew Members

This Notice supersedes Notice M.354
(DRIKOLD, CARDICE, DRY ICE, ETC.).

It is sometimes the practice to use solid carbon dioxide (Drikold, Cardice, Dry Ice, etc.) as an emergency refrigerant in defective cold chambers, and in the packaging of containers holding deep frozen food supplies such as ice cream, fish, etc. so as to retain their hard frozen condition during transit to ships.

The following precautions should be taken when solid carbon dioxide is used:-

1. Gloves or similar protection should always be worn when handling solid carbon dioxide to prevent blistering of the skin.
2. When solid carbon dioxide has been used as a refrigerant in a cold chamber, a high concentration of heavy carbon dioxide gas may be built up, and care should always be taken to open the door of such a chamber for some moments before entering, and to leave it open while inside.
3. Because of the heaviness of carbon dioxide compared with air, pockets of the gas are liable to collect in poorly ventilated spaces creating an unsafe atmosphere. As the gas does not diffuse away readily, special care should be taken to ventilate such confined spaces thoroughly before entering them.
4. The "Code of safe working practices for merchant seamen", copies of which should be on board all ships, contains detailed advice on entering enclosed or confined spaces. The warnings contained in the code should be heeded and the recommended procedures followed.

Merchant Shipping Notice No. M.984

USE OF LIQUEFIED PETROLEUM GAS (LPG) IN DOMESTIC INSTALLATIONS AND APPLIANCES ON SHIPS, FISHING VESSELS, BARGES, LAUNCHES AND PLEASURE CRAFT

EXPLOSIONS, FIRES AND ACCIDENTS RESULTING FROM LEAKAGE OF GAS

Notice to Shipbuilder, Owners, Masters, Skippers, Officers and Seamen of Merchant Ships and Fishing Vessels, Owners and Builders of pleasure craft and to other users of marine craft

This Notice supersedes Notice M.603

In view of the consiaerable use on smaller cargo ships, fishing vessels, tugs, barges, launches and pleasure craft of bottled hydrocarbon gases for cooking, water and space heating, refrigerators, etc., the Department wishes to draw attention to the possible dangers which may accompany their use and to the need for installations to comply at least with the requirements of the British Standard Institution publication BS 5482: Part 3: 1979—The code of practice for domestic butane and propane gas-burning installations; Part 3—Installations in boats, yachts and other vessels. Individual appliances and fittings should comply with the relevant British Standard Specifications listed in BS 5482: Part 3: 1979, some of which are given at Appendix 1.

2. The possible dangers associated with the misuse of such installations include fire, explosion and asphixiation due to the leakage of gas from appliances, storage containers or defective fittings or due to an accumulation of gas following flame failure. Incidents may result in loss of life and sometimes cause serious material damage. The siting of gas consuming appliances and storage containers and the provision of adequate ventilation of the spaces containing them are *consequently most important.*

3. In addition to the risk of asphyxiation should the leakage or accumulation of gas occur in an enclosed space, there is also the risk of carbon monoxide poisoning when the apppliance is in use. It is dangerous to sleep in spaces where gas-consuming open-flame appliances are left burning and it follows that heaters without flues should not be sited in areas designed as sleeping quarters or in unventilated spaces communicating directly with such areas.

4. Furthermore, open-flame heaters and gas refrigerators with non-enclosed burners may be present a serious hazard from the fire and explosion aspects and if possible, there use should be avoided.

5. In the United Kingdom the gases most commonly used for domestic Liquefied Petroleum Gas (LPG) installations in ships are butane or propane conforming to BS 4250—Commercial butane and propane. A stenching agent is added to enable the presence of gas to be detected by smell even when its concentration in air is below its lower limit of flammability. Trade names and the suppliers of some of these gases are given in Appendix 3.

6. It is important to remember with LPG installations that the gases, although heavier than air, if released, may travel some distance tending to fall to the bottom of a compartment. Here they diffuse and may form an explosive mixture with air, as in the case of petrol vapours.

7. A frequent cause of incidents involving LPG Installations is the use of unsuitable fittings or the replacement of items such as flexible hoses with temporary rubber or plastic tubing. It is essential that any repair or replacement part is in accordance with the original specification of the equipment as detailed in BS 5482: Part 3: 1979.

8. In view of the elements of danger in the use of LPG installations a warning notice in red should be displayed adjacent to each appliance to read as follows:

WARNING
1. **DO NOT LIGHT IF LEAKAGE IS SUSPECTED.**
2. **BEWARE OF ANY UNUSUAL SMELL AS THIS MAY INDICATE LEAKAGE FROM THE APPLIANCE.**
3. **DO NOT CHECK FOR LEAKS WITH A NAKED FLAME.**
4. **MAINTAIN GOOD VENTILATION AT ALL TIMES.**

9. In conjunction with any LPG Installation the provision of an automatic gas detection and alarm system of a reliable type is strongly recommended and is absolutely necessary when a cooking or other gas consuming appliance is fitted in sleeping or other spaces below decks. It is essential that any electrical equipment asssociated with the gas detection and alarm system should be certified as being flame-proof or intrinsically safe for the gas being used.

10. As expressed above LPG installations should at least comply with the requirements of BS 5482: Part 3: 1979 but the Department also wishes to stress the importance of obtaining expert advice regarding the fitting of LPG Installations and of the need to ensure that such installations and associated alarm systems receive adequate (and expert) maintenance in service.

11. BS 5482: Part 3: 1979 deals very fully with all aspects of LPG Installations but some general comments are made in Appendix 2 as all users of such installations may not have access to this publication or to the selection of individal specifications listed at Appendix 1.

12. LPG Installations in mechanically propelled sea-going fishing vessels registered in the United Kingdom need to comply, as appropriate, with the requirements of Rules 34 and 61 of the Fishing Vessels (Safety Provisions) Rules 1975. However the warnings detailed above are applicable to all fishing vessels and such installations not directly covered by the Fishing Vessel Rules should be in accordance with the recommendations of this Notice.

13. Attention is also drawn to the requirements of Regulation 6(6) of the Merchant Shipping (Crew Accommodation) Regulations 1978.

APPENDIX 1

SELECTION OF RELEVANT BRITISH STANDARD SPECIFICATIONS

BS 2491   Domestic cooking appliances for use with liquefied petroleum gases.

BS 2773	Domestic single room space heating appliances for use with liquefied petroleum gases.
BS 2871	Copper and copper alloys. Tubes. Part 1. Copper tubes for water, gas and sanitation. Part 2. Tubes for general purposes.
BS 2883	Domestic instantaneous and storage water heaters for use with liquefied petroleum gases.
BS 3016	Pressure regulators and automatic change over devices for use with liquefied petroleum gases.
BS 3212	Flexible rubber tubing and hose (including connections where fitted and safety recommendations) for use in LPG vapour phase and LPG/air installations.
BS 3605	Seamless and welded austenitic stainless steel pipes and tubes for pressure purposes.
BS 4104	Catering equipment burning liquefied petroleum gases.
BS 4368	Carbon and stainless steel compression couplings for tubes. Part 1. Heavy series. Part 3. Light series (metric).
BS 5045	Transportable gas containers. Part 2. Steel containers up to 130 litres water capacity with welded seams.
BS 5314	Specification for gas heater catering equipment.
BS 5386	Specification for gas heating appliances. Part 1. Gas burning appliances for instantaneous production of hot water for domestic use.

APPENDIX 2

GENERAL COMMENTS ON LPG INSTALLATIONS

1. Stowage of gas containers

Wherever possible gas containers should be stowed on the open deck or in a well-ventilated compartment on deck so that any gas which may leak can disperse rapidly. Where deck stowage is impracticable and the containers have to be stowed in a compartment below deck, such a space should be adequately ventilated to a safe place and any electrical equipment in the space should be of flame-proof construction. In all cases stowage should be such that the containers are positively restrained against movement, preferably in secure mountings specially designed for the purpose. On multiple container installations a non-return valve should be placed in the supply line near to the stop valve on each container. If a change-over device is used it must be provided with non-return valves to isolate any depleted container. Where more than one container can supply a system it should not be put into use with a container removed. Where stowage below deck or use of appliances in accommodation is unavoidable, an added precaution is the provision of remote closure of the main gas supply from the containers. Containers not in use or not being fitted into an installation should always have the protecting cap in place over the container valve. Containers should never be lifted by means of a rope around the valve.

2. **Stowage of spare and empty gas containers**
It is important that the stowage of spare and empty gas containers receive the same consideration as the positioning of operating containers, particularly with regard to ventilation and electrical equipment should the spare containers be stowed below decks.

3. **Automatic Safety Gas Cut-off Devices**
A device should be fitted in the supply pipe from the gas container to the consuming appliances which will shut off the gas automatically in the event of loss of pressure in the supply line, e.g. should a connecting pipe fracture. The device should be of a type which requires deliberate manual operation to re-set it to restore the gas supply. It is strongly recommended that all gas consuming devices should be fitted, where practicable, with an automatic shut-off device which operates in the event of flame failure.

4. **Open-flame heaters and gas-refrigerators**
Where such appliances are installed, they should be well secured so as to avoid movement and be preferably of a type where the gas flames are isolated in a totally enclosed shield, arranged in such a way that the air supply and combustion gas outlets are piped to the open air. However, in the case of refrigerators where the burners are fitted with flame arrestor gauzes, shielding of the gas flame may be an optional feature. Refrigerators should be fitted with a flame failure device and flueless heaters should be selected only if fitted with atmosphere-sensitive cut-off devices to shut-off the gas supply at a CO_2 concentration of not more than 1·5 per cent. by volume. Heaters of the catalytic type should not be used.

5. **Fittings and Pipework**
Solid drawn copper alloy fittings or stainless steel tube with appropriate compression or screwed fittings are recommended for general use for pipework for LPG installations. Aluminium or steel tubing and any materials having a low melting point such as rubber or plastic should not be used. Lengths of flexible piping (if required for flexible connections) should be kept as short as possible, be protected from inadvertent damage and comply with the appropriate British Standard.

6. **Ventilating Arrangements**
(*a*) It is highly desirable that compartments containing a gas-consuming appliance should not have access doors or openings direct to accommodation spaces or their passageways, but where this is impracticable it is advisable that mechanical exhaust ventilation trunked to within 12 in. of the floor adjacent to the appliances and adequate inlet ventilation be provided.
(*b*) Compartments containing a gas-consuming appliance which are situated upon an open deck with direct access to the adjacent deck and with no opening direct to accommodation spaces or their passageways should also be adequately ventilated, preferably by mechanical means.
(*c*) In pleasure craft and in some small ships where it may be impracticable to provide the mechanical ventilation referred to in sub-paragraphs (*a*) and

(b) above, all compartments where gas-consuming devices are used should have adequate natural ventilation of a type which cannot readily be closed and which will prevent a dangerous accumulation of gas. The ventilation should provide for extraction of any gas which might leak from the system, as well as providing a fresh air supply. Since the gas which is heavier than air, tends to fall to the lowest level, exhaust ventilation openings should be led from a position low in the space. Such ventilation might be provided by wind-actuated self-trimming cowls or rotary exhauster heads.

(d) When mechanical ventilation is fitted to any space in which gas containers or gas-consuming appliances are situated, the materials and design should be such as will eliminate incendive sparking due to friction or impact of the fan impeller with its casing. Electric motors driving fans should be situated outside the space and also, whenever practicable, outside the ventilation trunking and clear of outlets, but suitably certified flame-proof motors should be used if this cannot be achieved. Ventilator outlets should be in a safe area free from ignition hazard. Ventilation systems serving spaces containing storage containers or gas-consuming appliancs should be separate from any other ventilation system. Mechanical exhaust ventilation trunking should be led down to the lower part of the space adjacent to the appliance.

(e) Notwithstanding (b) above. Regulation 32(11) of part I of the Merchant Shipping (Crew Accommodation) Regulations 1978 requires mechanical exhaust ventilation to be provided for galleys in any ship over 1,000 tons gross and Regulation 25(9) of Schedule 6 to the same Regulations (existing ships) requires mechanical exhaust ventilation in any galley.

(f) In cases where loss of life has occurred due to asphyxiation or carbon monoxide poisoning the ventilation system has been found to be deficient because ventilators have been interfered with or neglected. It is not unusual to find ventilators deliberately blocked, and butterfly and sliding ventilators have been found to be in the closed position and immovable. The importance of adequate ventilation of spaces containing gas consuming appliances cannot be too strongly emphasised and on no account must a ventilation system be interfered with so as to prevent it functioning correctly.

(g) Whilst adequate ventilation is prerequisite for safety, consideration should be given to the siting of gas-consuming appliances in relation to the ventilation system such that air turbulance does not bring about the extinction of unshielded gas flames and thus permit the escape of gas.

7. Gas Detection

Suitable means of detecting the leakage of gas should preferably be provided in each compartment containing a gas-consuming appliance and where this is a detector, it should be generally be securely fixed in the lower part of the compartment in the vicinity of the gas-consuming appliance. Any gas detector should preferably be of a type which will be actuated promptly and automatically by the presence of a gas concentration in air not greater than 0·5 per cent (representing approximately 25 per cent of the lower explosive limit) and should incorporate an audible and a visible alarm, although on small craft a portable manually operated detector may be used. Where electrical detection equipment is fitted it is essential that it should be certified as being flame-proof or intrinsically safe for the gas being used. In all cases where detection and alarm equipment are used, the alarm unit and indicating panel should be situated outside the space containing the gas storage and consuming appliances.

Similar provision for automatic gas detection and alarm should also be made in small vessels, such as pleasure craft and barges, if a cooking or other gas-consuming appliance is fitted in sleeping or messing spaces below deck.

Detectors can be rendered unsafe for use in explosive atmospheres by inexpert servicing, particularly in respect of arrangements for sealing off the detection chamber. Any maintenance should therefore be carried out by persons competent to do so or by replacement of the detection unit.

In all cases the arrangements should be such that detection devices can be tested frequently whilst the craft is in service.

8. Emergency Action

A suitable notice detailing action to be taken when an alarm is given by the gas detection system should be displayed on board the craft. In addition, the information given should include the following:

(*a*) the need to be always alert for gas leakage:

(*b*) when leakage is suspected all gas-consuming appliances should be shut off at the main supply from the container and no smoking should be permitted until it is safe to do so. **NAKED LIGHTS SHOULD NEVER BE USED AS A MEANS OF LOCATING LEAKS:**

(*c*) the correct use and maintenance of fire extinguishing appliances of which an adequate number should always be carried;

(*d*) the need for users to be fully aware of the contents of the consumer instructions and emergency procedures issued in accordance with clause 22 of BS 5482: Part 3: 1979.

APPENXIX 3

SUPPLIERS OF LPG GASES USED IN THE SHIPPING INDUSTRY

Supplier	*Gas*
1. **Propane**	
Air Products Ltd.	Propane
British Oxygen Co. Ltd.	Propane
BP Oil Ltd.	BP Gas—Propane
Calor Gas Ltd.	Calor Propane
Calor Kosangas (Northern Ireland) Ltd.	Propane
Shell UK Oil	Propagas
2. **Butane**	
BP Oil Ltd.	BP Gas—Butane
Calor Gas Ltd.	Calor Gas
Calor Kosangas (Northern Ireland) Ltd.	Butane
Shell UK Oil	Butagas

Merchant Shipping Notice No. M.1136

FIRES INVOLVING OIL FIRED APPLIANCES

Notice to Shipowners, Shipbuilders, Ship Managers, Masters and Crew Members of Small Cargo Ships and Similar Vessels and to Owners, Builders, Managers, Skippers and Crew of Fishing Vessels.

This Notice supersedes Notice M.851

1. There have recently been several accommodation fires on small fishing vessels and subsequent investigations found that the fires had been caused by oil fired appliances which had neither been installed nor operated in accordance with the manufacturers instructions. In several cases the appliance had not been fitted with any device by which the fuel supply could be shut off in the event of fire. In one particular case resulting in loss of life, in which a fire had been caused by overheating of the uptake, the safety features designed to shut down the heater in the event of fire had been removed.

2. Any oil fired appliance, whether it is an accommodation space heater, galley stove or similar appliance should be installed and operated in accordance with the manufacturer's recommendations including those for siting the fuel supply and making arrangements for the flue from the appliance. In addition, the fuel shut off valve to any oil fired appliance should be readily identifiable and sufficiently remote from the appliance so as to be accessible in the event of a fire involving the appliance.

3. The appliance should be regularly and frequently examined to ensure that it is maintained in a condition that will enable it to operate properly. Particular attention should be given to the controls for regulating the oil supply, to the means fitted to shut down the heater in the event of fire and to the security of the oil connections.

4. When installing an oil fired appliance in accommodation or other enclosed spaces particular attention must be given to the provision of adequate ventilation and again the manfacturers recommendations should be followed. The ventilation arrangements should be regularly and frequently examined to ensure that they are not obstructed or have not been interfered with and that any moving parts can be operated satisfactorily. Oil fired heating boilers should operate in forced draught with suitable safety features to cater for flame failure. It is strongly recommended that such boilers should not be operated using natural draught as this may result in a blow back in variable wind conditions.

5. The attention of Shipowners and Masters of Cargo Ships is drawn to the provisions of Regulations 14(3) and 14(5) of Merchant Shipping (Crew Accommodation) Regulations 1978 which indicate that:

  (a) a heating appliance shall be provided with a means of turning it on or off or controlling the heat without the use of a tool or key and such means shall be, wherever reasonably practicable, within the space in which the appliance is fitted;

  (b) the heating appliance shall not be affected by the use or non-use of the ships machinery, calorifiers or cooking appliances; and

  (c) the heating appliance shall be constructed and installed and if necessary shielded so as to avoid risk of fire or of danger or discomfort to the crew.

Regulation 38(1) of the above Regulations requires that all equipment and installations shall be maintained in good working order, and Regulation 38(2) requires a Master to inspect crew accommodation and record defects in the Official Log Book.

6. The attention of Owners and Skippers of Fishing Vessels is drawn to the provisions of Regulations 14(3) and 14(5) and Regulations 35(1) and 35(2) of the Merchant Shipping (Crew Accommodation) (Fishing Vessels) Regulations 1975 which are similar to those of the Regulations referred to in paragraph 5 above.

7. The Owners and Skippers of Fishing vessels are reminded that where the Fishing Vessels (Safety Provisions) Rules 1975 apply, the installation of space heaters and cooking stoves shall be in accordance with Rules 32 and 61(2) of these Rules. The relevant requirements of these Rules are as follows:
   (a) When a heating or cooking appliance is supplied with fuel from an oil tank, the tank shall be situated outside the space containing the appliance and the oil supply shall be capable of being controlled from outside that space.
   (b) Appliances using oil fuel having a flash point of less than 60°C (Closed Test) shall not be fitted.
   (c) Means shall be provided to shut off the fuel supply automatically at the appliance in the event of a fire or failure of the air supply. This means shall require manual resetting in order to restore the fuel supply.
   (d) The oil tank supplying the appliance shall be provided with an air pipe leading to a position in the open air where there is no danger of fire or explosion from the oil vapours. The open end of the pipe shall be fitted with a detachable wire gauze diaphragm.
   (e) Means shall be provided for filling the oil tank and for preventing overpressure.
   (f) Appliances shall be secured in position and their exhausts and the surrounding structure shall be adequately protected against fire. Exhausts shall be provided with ready means of cleaning. Dampers fitted in exhausts shall provide an adequate flow of air when in the closed position.
   (g) Ventilators which are used to provide an adequate supply of air to a heating or cooking appliance shall not be capable of being closed.

8. It is strongly recommended that the installation of space heaters and cooking stoves in Fishing vessels of less than 12 metres in length to which Rules 32 and 61(2) of the Fishing Vessels (Safety Provisions) Rules 1975 do not apply should be generally in accordance with the requirements of these Rules indicated in Paragraph 7 above.

9. Combustible bulkheads, ceilings, linings and furniture should be protected from adjacent oil fired appliances such as galley ranges, space heaters etc. by non-combustible board type materials not containing asbestos.

10. Oil fired appliances such as galley ranges, space heaters etc. should not be used for drying clothing, linen etc. because of the danger that such clothing, linen etc. may interfere with the ventilation arrangements of the appliance or fall on to the appliance and catch fire. Curtains and other hanging textile materials should not overhang or be fitted sufficiently close to an oil fired appliance where there is any danger of them coming into contact with the appliance.

Merchant Shipping Notice No. M.830

ACCIDENTS WHEN USING POWER OPERATED WATERTIGHT DOORS

Notice to Ship Owners, Masters, Officers and Crew

1. In recent years there have been a number of avoidable accidents involving people being seriously injured or killed by the closing of power operated watertight doors.

2. Most of these acidents have occurred when crew members were using the local controls to pass through the watertight doors which had been closed from the bridge control station; under these circumstances if the local control is released the door would automatically close with a force sufficient to crush anybody caught in its path. It is absolutely essential therefore that the local controls, which are provided on both sides of the bulkhead, are held in the "open" position while passing through the door and this is easily achieved by first opening the door using the nearside control with one hand, then reaching through the opening to the other control and maintaining the door open until transit is complete. A man by himself needs both hands to operate the controls and should never attempt to carry any load through unassisted.

3. It is seldom that there are any witnesses to or survivors from watertight door accidents, but investigations of recent casualties indicate that unfamiliarity with, and consequent inattention to, operating instructions had probably been the prime cause. It would seem that the victims have not always followed the correct operating procedure but have opened the door and then attempted to pass through quickly while the door was closing: with the result that they have either slipped, or have allowed insufficient time to pass through, and have been consequently trapped by the door. When the door is closing an audible warning is given, during which time no attempt should be made to pass through the door.

4. Many proposals for making watertight doors safer have been investigated but unfortunately all (in certain circumstances) could hinder closing in an emergency, thus possibly hazarding the ship and all on board. Furthermore there is no record of people being injured by watertight doors when properly operated. If further accidents are to be avoided it is essential that all crew members are well trained in the correct operating procedure and that permanent notices clearly stating the correct operating procedures are prominently displayed on both sides of each door. The crushing power of watertight doors must be fully appreciated by those using them; this power, together with expeditious closing, is necessary to ensure that their primary purpose is fulfilled.

5. It has sometimes been found, when only one door closing warning bell has been fitted (normally giving adequate warning), that it cannot always be clearly heard on both sides of the opening door, particularly in noisy machinery spaces. In such a case a bell (or other suitable audible signal) should be fitted on each side of, and adjacent to, the watertight door concerned.

6. All powered operated watertight doors are designed to ensure maximum safety of the ship and to persons using them, but safe transit through a door which has been closed from the bridge control station requires that the operating instructions be strictly observed.

CHAPTER 8.
ROPE AND ROPEWORK.

TYPES OF ROPE.

Rope may be constructed from natural fibres such as cotton, coir, hemp, manila and sisal or from synthetic fibres such as polyamide (nylon), polyester (terylene), polythene and polyproplene or a mixture of some of these synthetic fibres. It may also be made from drawn strands of steel wire or a mixture of wire and either natural or synthetic fibres. Wire ropes for marine use are normally galvanised and all ropes for marine use are normally laid up right handed (anti-clockwise) and supplied in coils of 120 fathoms (220 m), unless specially ordered otherwise.

Natural fibres.

Cotton rope is mainly used on yachts, being soft and pliable.
Hemp rope is used for small lines and edging sails.
Coir rope is both buoyant and elastic but lacking in strength, normally used for guess warps and when attached to wire rope tails, for mooring springs and tow ropes.
Sisal rope is in general use for gantlines, lashings and moorings.
Manila rope is in general use for Lifeboat falls, cargo handling, lashings and moorings.
Coir, hemp and cotton ropes are no longer much in demand, having been largely superseded by synthetic fibre ropes.

Synthetic fibres.

Polyamide (Nylon) rope, the most elastic and strongest of all synthetic fibre ropes, resists alkalis, oils, organic solvents and rot. Melting point 250°C. Mainly used for springs and tow ropes, also used by stevedores for cargo work.
Polyester (Terylene) rope with a melting point of 260°C has the highest resistance to fusing of all synthetic fibres. Resists acids, oils, organic solvents, bleaching agents and rot. Mainly used on yachts.
Polythene rope with a melting point of 135°C has the least resistance to fusing of all synthetic fibres. Buoyant, it resists alkalis, oils, bleaching agents and rot. Used for loglines and halyards.
Polypropylene rope is made in three types, "Fibrefilm", "monofilament", and "staple". The main difference being in the elastic property of the rope, the least elastic and lightest of the synthetic fibre ropes, it has a melting point of 165°C and is buoyant. Resists acids, alkalis, oils and rot. Main use is for mooring lines. Only spun staple polypropylene should be used for gantlines.

Preformed Galvanised Steel Wire rope.

Non-flexible galvanised steel wire rope (Iron wire) is manufactured for use as standing rigging (shrouds, stays, etc.), and will have six strands laid up right handed on a wire core.
Seizing wire as used for wire seizings will have six wire threads laid up right handed on a wire core.
Flexible steel wire rope used for running rigging and moorings, and when attached to fibre rope as tails for springs and towing. May have 6, 12, 17 or 18 strands laid up right handed on a fibre core. The wire threads in each strand will

also be laid up on fibre cores. Generally speaking, cargo runners and moorings will contain 6 strands while deck cranes will have more, lifeboat falls are often 18 strand.

The greater the number of strands and the greater the number of wire threads in each strand, the greater the flexibility of the rope. Wire ropes that are required to work cargo are to be discarded if in any length of eight diameters, the total number of visible broken wires exceeds ten per cent of the total number of wires, or if the wire shows signs of excessive wear or corrosion or other serious defect. No chain or wire rope shall be used when there is a knot tied in any part thereof.

Mixed wire and fibre ropes.
Combined rope four or six wire strands laid up right handed over a natural fibre (sisal) core. Each wire strand is covered with natural fibre (sisal) yarns. Combined rope is mainly used on fishing vessels and cable ships.
Spring lay rope six main strands laid up left handed (cable laid) on a fibre core. Each main strand comprises three wire strands and three fibre strands laid up right handed on a fibre core. The fibre may be either treated (tarred) sisal or polypropylene "fibrefilm". Springlay ropes are normally supplied in 90 fathom (165 m) coils and are used as mooring lines and backsprings.

Rope construction.
Normally the threads in each strand of a rope will be laid up left handed and the strands will be laid up right handed. However, non-kinkable rope such as may be used for lifeboat falls has both the threads and the strands laid up right handed.
Hawser laid rope contains three strands laid up right handed.
Cable laid or water laid rope contains three hawser laid ropes laid up left handed. Synthetic fibre and wire cable laid ropes, are laid up on a core.
Shroud laid rope contains four strands laid up right handed on a core.
Plaited rope is supplied where the rope is subjected to twisting viz., log lines and some halyards.
Squareline and multiplait ropes contain eight strands. Two pairs are laid up right handed and two pairs are laid up left handed, so forming a plait.

Six stranded synthetic fibre rope and all wire ropes are laid up on a core.

Originally in the manufacture of steel wire rope the strands and threads were forcibly held in place, this could be seen by cutting the rope when all the strands and wire threads immediately flew apart. Preforming the wires and strands prevents this and increases the strength. The strands all lie naturally in their true positions without stress being applied. Apart from the fact that this gives the rope a greater breaking strain and longer life, should one or more wire threads break, the ends will lie flat with the wire to the great advantage of any person handling the wire and the risk of a badly torn hand is considerably reduced.

Breaking strains and Safe working loads.
To find the breaking strain or the safe working load of any rope, the manufacturer's tables must be consulted. However, in an emergency it is possible to obtain a rough guide to the breaking strain of good quality manila rope with the formula $\frac{C^2}{3}$ (the circumference in inches multiplied by itself and the answer divided by three will give the approximate breaking strain in tons).

A rough guide to the breaking strain of flexible steel wire rope may be obtained with the formula $2C^2$ (the circumference in inches multiplied by itself

ROPE CONSTRUCTION.

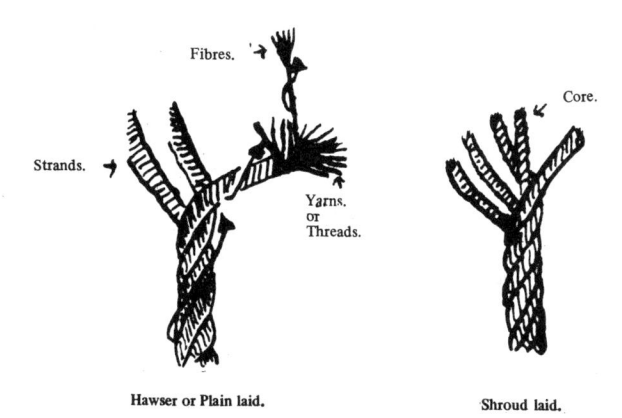

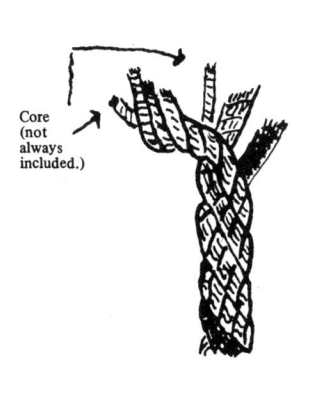

Hawser or Plain laid. (Fibres, Strands, Yarns or Threads.)

Shroud laid. (Core.)

Cable or Water laid rope. (Core (not always included.))

Multiplait or Squareline rope.

Steel wire rope. (Strands, Core, Wires, Core.)

and the answer doubled will give the approximate breaking strain in tons). The safe working load of manila rope may be found by dividing the breaking strain of the rope by six. The safe working load of flexible steel wire rope may be found by dividing the breaking strain by five.

Fibre ropes should not be worked through a block whose cheek length is less than three times the circumference of the rope.

Wire ropes should not be worked over a sheave whose diameter is less than sixteen times the diameter of the rope, or five times the circumference of the rope.

Under no circumstances will the safe working load marked on any tag attached to any rope or coil of rope or the safe working load marked on any derrick, block, shackle or other tackle, ever be exceeded. Where a union purchase is operated, the safe working union purchase (u) load as marked on the derrick must not be exceeded and will be about one third of the normal safe working load.

Care in the use of rope.

To open a coil of any rope, a turntable should always be used, the rope being taken from the outside of the coil and recoiled clockwise. Great care must be taken to avoid any kinks getting into the rope. Fibre ropes are to be coiled down on duckboards, (small lines may be hung up in a ventilated but dry place.) Wire ropes should be stowed on reels. Never leave a rope coiled on a bare deck because this prevents moisture from draining which in turn will cause rot or corrosion. Try not to drag rope over the deck, because it will pick up grit that may later cause serious damage to the centre of the rope from abrasion as the grit works in.

When no turntable is available, fibre ropes may be uncoiled by taking the end of the rope from the interior of the coil left-handed or anti-clockwise (this will usually be from the bottom of the unopened coil) and recoil right-handed or clockwise. Wire rope may be uncoiled by rolling the coil along the deck and then recoiling clockwise, or better still be wound onto a reel. No attempt should be made to uncoil a wire rope by taking the end from the centre of the coil.

A fibre rope that is full of kinks or turns may have the turns removed by being thoroughfooted. To thoroughfoot a rope containing left hand turns, coil down left handed, then dip the end through the coil and haul out. To thoroughfoot a coil containing right hand turns, coil down right handed, then dip the end through the coil and haul out. An alternative method of removing turns is to keep one end of the rope on deck while the remainder of the rope is thrown down an empty hold. The end of the rope is then taken to a winch drum end and the rope hove out of the hold. Someone should be stationed at the bottom of the hold to twist the rope and work the turns back as they are hove up the hold. The rope being coiled down right handed as it comes off the winch drum end. Do not throw the end of the rope off the winch drum end, allow the end to go round and work the last of the turns out.

Fibre and wire ropes should whenever possible be kept separate, try not to put wire and fibre ropes on the same pair of bitts or bollard or through the same fairlead, in any case the wire rope must never be allowed to cross the fibre rope, for this causes considerable damage to the fibre rope.

Unless otherwise stated, all fibre ropes are liable to deteriorate when static, exposed to strong sunlight, chemical fumes, heat, sparks and by abrasion, or when contaminated by acids, alkalis, bleaching agents, oils and organic solvents.

Opening up a hank of seaming twine to make a skein.

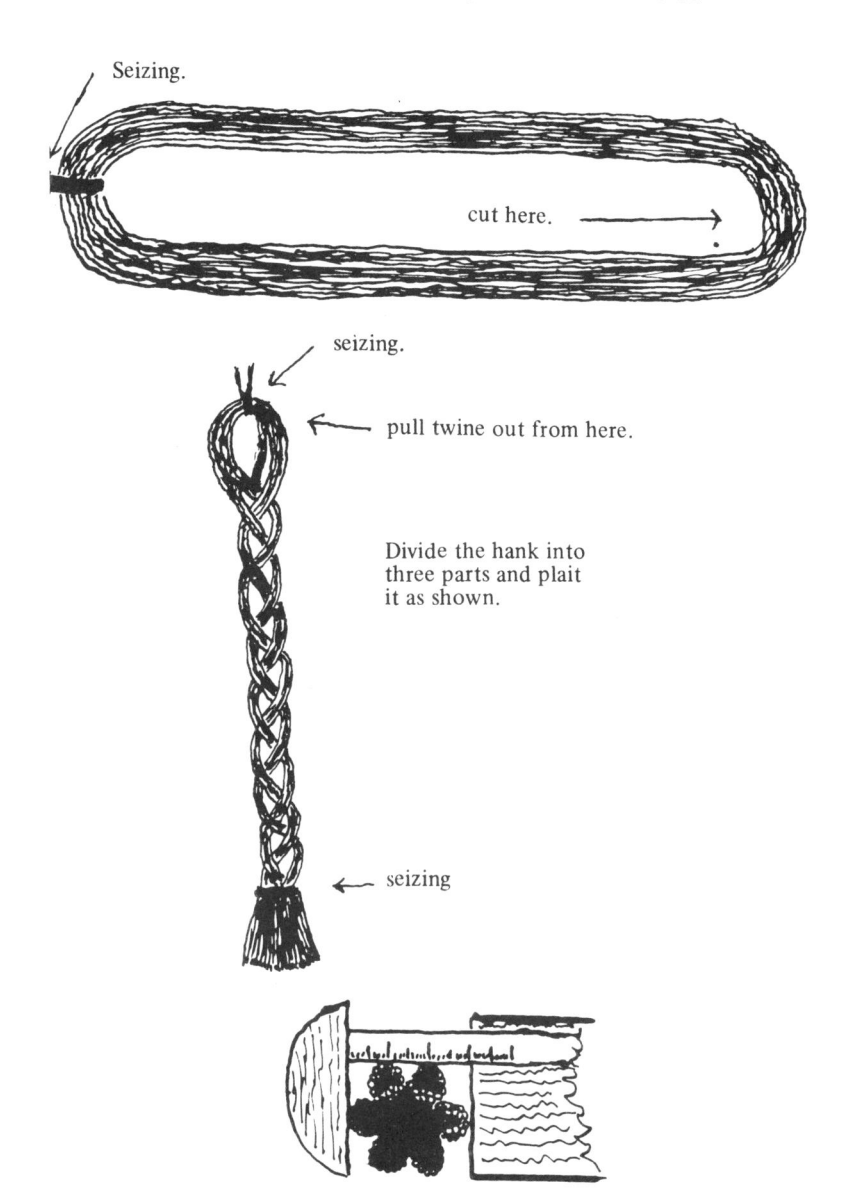

Seizing.

cut here.

seizing.

pull twine out from here.

Divide the hank into three parts and plait it as shown.

seizing

Measuring the circumference in inches or the diameter in millimetres by means of a rope guage.

Care must be taken never to allow any rope to come in contact with white spirit, rust remover, wet paint, coal tar, xylene, metacresol and similar substances. In the event of contamination, the rope should be immediately washed with fresh water and closely inspected for damage. Wire rope is mainly subject to deterioration from moisture and abrasion and should be kept well oiled. Natural fibre ropes are particularly subject to damage from mildew and rot and must not be stowed away wet.

Rope should never be stored in the vicinity of steam pipes or boilers. Fibre ropes are to be coiled down right handed on duck boards or hung up in a sheltered and dry but well ventilated position, natural fibre ropes should not be stored under conditions where there is any danger of either mildew or fungous growth. Wire ropes are to be kept wound on reels or winch barrels and protected from weather by canvas covers. Mooring wire ropes stowed on reels are to have their ends made fast to the reel with a light lashing, so that if the wire runs out, the lashing will part and not take the reel. When stowed on deck, fibre ropes should be covered with tarpaulin canvas in order to exclude damage from weather and strong sunlight. Under arctic conditions nautral fibre ropes should be protected from rain, spray, snow, frost and ice. All fairleads are to be kept free of rust and well oiled in order to avoid both friction and abrasion. Winch drum ends and capstans should be both smooth and free from rust.

Every rule has an exception. Log and lead lines are always to be coiled left handed or anti-clockwise.

Before use all fibre rope gantlines, lizards and safety harness lines should be tested to at least four times the weight they may be expected to support. This is to insure there is no unapparent damage.

Never leave the end of a rope to fray. Put a whipping on.

Never use a rope without previous careful examination of its whole length.

TYPES OF CORDAGE.

Seaming twine. Three ply hemp used for sewing canvas and whippings. Supplied in ½ lb. (0.28kg) hanks.

Roping twine. Five ply hemp used for sewing canvas to rope. Supplied in ½ lb. (0.28kg) hanks.

Marline. Tarred two ply hemp used for serving wire rope and seizings. Supplied in 30 fthm. (55m) balls.

Spunyarn. Tarred three ply hemp used for worming, serving and seizings. Supplied in 30 fthm. (55m) balls.

Boat lacing. Hawser laid hemp or plaited Polypropylene, used for fastening boat covers and awnings. Supplied in two weights in hanks of 30 & 60 fthms. (55 & 110m).

Signal halyard. Hawser or shroud laid hemp used for flag hoists. Supplied in hanks of 30 or 60 fthms. (55 & 110m). Plaited hemp or polythene. Supplied in coils of 40 fthms. (73m).

Log line. Plaited hemp or polythene. Supplied in coils of 40, 60 and 120 fthms. (73, 110 & 220m).

Lead line. Cable laid hemp used for hand and deep sea lead lines. Supplied in hanks of 30 fthms. (55m) and coils of 120 fthms. (220m).

Ratline. Hawser laid tarred hemp used for heaving lines and light lashings. (Originally used for rattling down the rigging). Supplied in coils of 120 fthms. (220m).

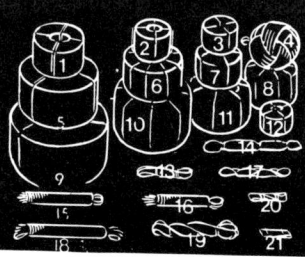

SUNDRY LINES

1. Hand Leadline
2. Plaited Halyard
3. Plain laid Logline
4. Fender
5. Hemp Rope
6. Pointline
7. Plaited Logline
8. Deep-sea Leadline
9. Ratline
10. Gaskin Yarn
11. Spunyarn
12. Plain Halyard
13. Houseline
14. Condenser Cord
15. Hambroline
16. Thermometer Cord
17. Roping Twine
18. Boat & Awning Lacings
19. Seaming Twine
20. Marline
21. Indicator Cord

Pointline.   Hawser laid sisal or manila used for general purposes. Supplied in coils of 120 fthms. (220m).
Boltrope.   Hawser laid tarred or untarred hemp or polyester used for edging sails or canvas. Measured by circumference or diameter in sizes from ½" (4mm) to 6" (48mm) and supplied in coils of 120 fthms. (220m).

Rope measurement.
Small natural fibre cordage is measured by the ply (number of threads) or by a number or by weight.
Small synthetic fibre cord is measured by a number.
Ratline and pointline are measured by the total number of threads contained in all the strands, i.e., 9, 12, 15, 18 and 21 threads.
Seizing wire is measured by the gauge (18 to 26) and supplied in seven pound (3.18kg) coils.
All other ropes are measured by the circumference in inches or by the diameter in millimetres. When measuring rope, if no rope gauge is available, measure round the circumference with a piece of twine and lay the length of twine against a ruler.
To turn circumference in inches into diameter in millimetres:—
    multiply by 8. i.e., 2 inch rope x 8 = 16mm rope.
To turn diameter in millimetres into circumference in inches:—
    divide by 8. i.e., 12mm rope ÷ 8 = 1½ inch rope.
Ropes may be marked by a coloured thread if this is required by the purchaser but there is no standard marking for any particular type of rope, except that preformed ungalvanised steel wire rope will be marked with a blue thread and polypropylene cordages exceeding 12 mm (1½ inch) will incorporate a tape throughout their length bearing at intervals the words "DOT accepted for LSA" if the cordage has been accepted as suitable for use with life-saving equipment. (M Notice No. 698).

VARIOUS ROPES.

Aerial downhaul.	Wire rope from truck to deck, used to hoist and lower W/T aerial.
Back spring.	A mooring rope, usually wire, leading aft from the bow or forward from the stern.
Backstay.	Strong wire rope used to help support a mast, always leads a little aft of the mast.
Bight.	Any part of a rope except the end.
Boat-rope.	A length of fibre rope made fast in the bows and hung overside for a small boat to catch, when the ship is under way.
Bowsing-in rope.	Several turns of fibre rope around a stanchion and a lifeboat fall. Used to bind the fall close to the ship's side, while the boat is lowered.
Breast rope.	A mooring line leading athwartships.
Bull rope.	Rope used for topping a derrick so that the topping lift can be shackled to a chain preventer.—Rope used for pulling an object into the square of the hatch.
Dummy gantline.	A short rope or chain rove through a sheave permanently to enable a gantline or runner to be drawn through when required.
End.	The free end of any rope.

Term	Definition
Fall.	A rope rove through one or more blocks to make a whip, purchase or tackle.
Gantline.	A fibre rope used aloft to lower a man in a bosun's chair or overside with a stage.
Guess warp.	A length of coir or polythene rope hung overside for the use of small boats when the ship is at anchor. (Guest Warp).
Halyard.	A fibre rope used to hoist a flag, sail, anchor ball, navigation light or other similar article.
Hauling part.	The part of a rope that comes out of a block, onto which a pulling motion is imparted to bring a purchase or tackle into use.
Hawser.	Any rope three inches (24mm) or more in circumference.
Head rope.	A mooring rope leading ahead from the bow.
Heaving line.	A length of ratline used when entering port, with the end thrown ashore it acts as a messenger to take the mooring lines. (about 15 to 20 fthms (27 to 37m) long).
Heel rope.	Wire rope used to lower a telescopic topmast.
Jumper stay.	A fore and aft wire rope from mast to mast or funnel. Originally used to work cargo but now used to carry flag halyards.
Lacing.	A light fibre line used to secure a boat cover or an awning.
Lanyard.	Short piece of fibre rope used to secure something.
Lashing.	Length of rope used to secure something.
Lifeline.	Light fibre rope becketed around a boat or buoyant apparatus for the support of persons in the water.—Heavier manila rope attached to the davit head span for the use of persons in a boat which is being either hoisted or lowered.
Lizard.	Length of fibre rope having a thimble eye spliced in one end, the other end being secured. A boat's painter or stage gantline may be rove through the thimble eye.
Man rope.	A fibre rope running through stanchions on a gangway or accommodation ladder. A rope stretched along a deck in bad weather or over a deck cargo, to provide a handhold.
Messenger.	A light line sent ashore or elsewhere to enable a heavier line to be hauled out.
Mooring line.	Any rope used to tie a ship to a quay or jetty.
Painter.	A fibre rope used for mooring a small boat.
Pendant.	A short length of wire rope having an eye in each end, used for hanging off a weight.
Preventer.	A rope made fast to a strong point and something moveable, as a second method of fastening which will hold the movable object, if the tackle holding the moveable object should break.—Preventer guys on derricks—preventer stays on masts.
Purchase.	Two blocks with a rope rove through them, also called a tackle. Used to increase the strength of the power applied to the hauling part.
Ridge rope.	Wire rope towards which an awning or dodger is stretched and laced onto.
Runner.	Wire rope used on a derrick for working cargo as a whip or in a purchase.

Shrouds.	Strong wire ropes that help to give thwartship support to a mast. Attached to the mast in pairs.
Snorter.	A short rope having an eye in each end. (Snotter).
Standing part.	The end of a rope which is made fast, or a part of the rope which is toward that end, or the part which is toward the main body of the rope.
Stay.	Strong wire rope that leads forward from a mast and helps to support the mast.—Strong wire ropes that lead in any direction and help support a funnel, samson post or other object.
Stern rope.	Mooring rope leading astern from the stern of the ship.
Stopper.	A short length of fibre rope or chain with an eye in one end in the case of a rope, or a small shackle in the case of a chain. Used to take the strain of a mooring rope or topping lift, whilst the rope is being moved from the winch to the bitts.
Strop.	A rope having its two ends spliced into each other.
Swifter.	Additional single shroud.
Tackle.	Quantity of rope with or without blocks, shackles, hooks etc., used for some purpose. Or a purchase.
Topping lift.	A wire rope used for topping a derrick or small mast.
Tow rope.	Heavy rope used for towing. Usually a fibre rope attached to a wire rope tail.
Triatic stay.	Jumper stay.
Tricing line.	Light line used to recover a fall from overside, sometimes attached to a small block through which the fall is free to run.
Tricing pendant.	Short length of wire rope having an eye in each end. Attached to gravity davits for the purpose of hanging off the boat to bring it alongside, when the boat is lowered on the listed side of a ship.
Warp.	Mooring line when in use to warp the ship.
Whip.	A fibre rope running through a gin block, used for working cargo etc., on a winch drum end.
Yard lift.	Wire rope fixed to the mast and supporting a yard arm.

KNOTS, BENDS AND HITCHES.

Reef knot.	Used to join together two ends of rope that are of equal texture and circumference. When properly made will not slip. UNSAFE when used with ropes of differing circumference or texture
Bowline.	A loop made on the end of a length of rope, the size of the loop remains static and will not draw tight.
Bowline on the bight	Two loops, which are adjustable for size with regard to each other, made in the bight of a rope. Used in lieu of a bosun's chair.
Timber hitch.	A loop made on the end of a length of rope and which will draw tight (note that the end is dogged with the lay). Used to lift sacks of grain etc., when used to lift timber an additional half hitch is added to slant the timber.

REEF KNOT.

BOWLINE.

TIMBER HITCH.

BOWLINE ON THE BIGHT.

TIMBER HITCH AND HALF HITCH

SINGLE SHEET BEND.

CLOVE HITCH.

COW HITCH.

DOUBLE SHEET BEND.

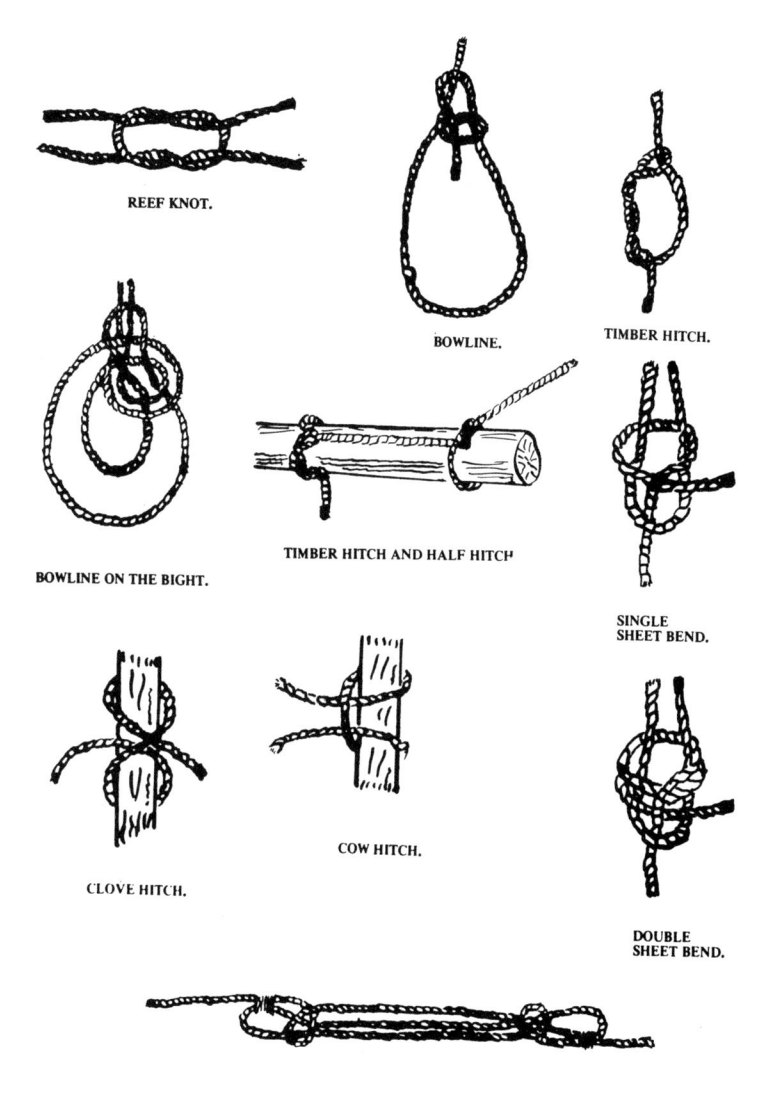

SHEEPSHANK.

Join the end of a rope to a soft eye with a sheet bend. Join the end of a rope to a hard (thimble) eye with a round turn and two half hitches.

ROLLING HITCH.

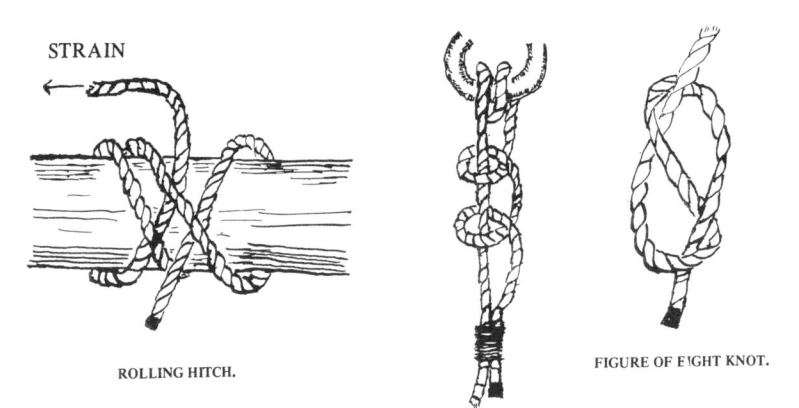

FIGURE OF EIGHT KNOT.

ROUND TURN AND TWO HALF HITCHES.

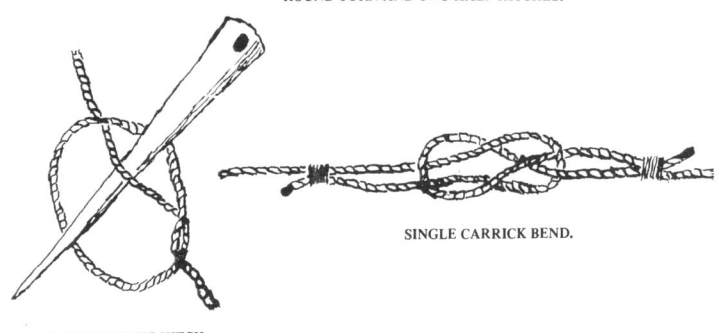

MARLINE SPIKE HITCH.

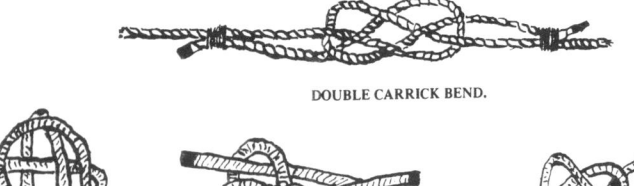

SINGLE CARRICK BEND.

DOUBLE CARRICK BEND.

CROWN KNOT.

WALL KNOT.

WALL AND CROWN.

Wall and crown:- Make a wall knot first and a crown on top.

BOSUN'S CHAIR WITH LOWERING HITCH.

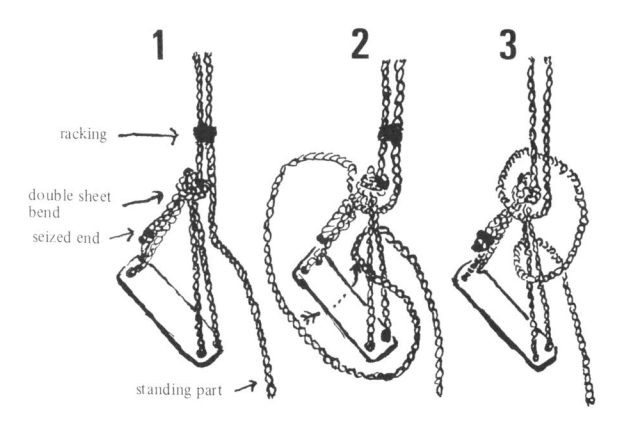

Labels on figures: racking; double sheet bend; seized end; standing part.

1. Gantline made fast to the bridle with a double sheet bend, having the end of the gantline seized to a leg of the bridle.
   The standing part of the gantline runs up through a sheave in the top mast and is then seized to its own part with a racking.

2. The bight of the gantline is now passed underneath the chair and up round the bridle.

3. The racking is let go and the gantline forms a cow hitch around the bridle.
   by easing the standing part of the gantline into the hitch, the occupant is able to lower himself.

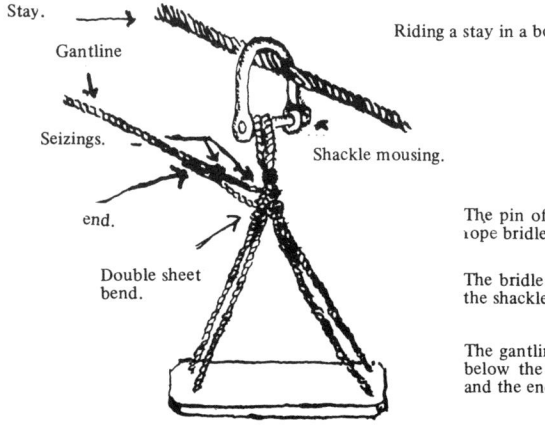

Riding a stay in a bosuns' chair.

Labels: Stay. Gantline. Seizings. end. Double sheet bend. Shackle mousing.

The pin of the shackle must go through the rope bridle of the chair and be moused.

The bridle is to be seized to form an eye for the shackle pin.

The gantline is to be made fast to the bridle below the seizing with a double sheet bend and the end seized.

Slinging a cask end up on a ropes end using an overhand knot and a reef knot.

Overhand knot.

Under no circumstances is a drum or cask ever to be lifted end up by means of a handle in the lid, except for the purpose of being carried by hand.

Slinging a cask end up using a strop and two half hitches.

Marrying two ropes with a palm and needle in order to lead a gantline over a sheave with a dummy gantline.

Single sheet bend.	Used to join together two ends of rope that are of unequal circumference or texture. When properly made will not slip. (Take the lighter rope through and around the bight of the heavier rope.)
Double sheet bend.	A more secure version of the single sheet bend. Always to be used when making fast a gantline to a bosun's chair.
Clove hitch.	Used to make the end of a rope fast to a fixed spar. Will pull out when made fast to a rotating spar.
Cow hitch.	Used with a chain stopper. Two half hitches, the second half hitch being reversed, also used as a lowering hitch on a bosun's chair. Unsuitable for securing an end.
Sheepshank.	Made in the bight of a rope to shorten it without cutting. Two loops are formed which must be seized to the standing parts to ensure that the hitch does not come adrift when slack. Chiefly used on the keel grab lines of lifeboats.
Rolling hitch.	Used to secure the end of a rope to a fixed spar. When properly made the rope will not slip along the spar in the direction of a strain imposed on the standing part.
Round turn and two half hitches	For securing a rope to a ringbolt. If the rope is liable to get wet, the half hitches should be separated and the end sized to the standing part in order to prevent the hitches jamming.
Figure of eight.	Placed in the end of a rope to act as a stopper and prevent the end running through a block or eye. Sometimes placed at intervals along the length of a lifeline to provide handholds.
Marline spike hitch.	Made in the bight of a line with a marline spike, it will not slip and allows leverage to be put on the marline spike to draw the line tight. Also used on the end of a stage as an alternative to a stage knot.
Wall and crown.	Placed at the end of a rope to provide a stopper knot and prevent fraying. Frequently used to prevent the end of a man rope passing through the eye of a stanchion. Known as a man rope knot when the ends are followed round a second time.
Carrick bend.	Another method of joining two ropes, when properly made will not slip and does not jam. Use a single carrick bend to make the end of a fibre rope fast to a soft eye in the end of a wire or fibre rope. The end MUST be seized. Use a double carrick bend to make the ends of two fibre ropes fast to each other when the ropes are expected to take a heavy strain and get wet. The ends MUST be seized.
Stage knot.	Used to secure each end of a stage when working overside or up a funnel. A marline spike hitch is an alternative that is sometimes used.
Lowering hitch.	Formed in the bight of a gantline, makes a cow hitch around the bridle of a bosun's chair, by means of which the operator may lower himself. Differently formed by means of turns around the end of a stage and half hitched to the leg, it allows the operators to lower themselves.

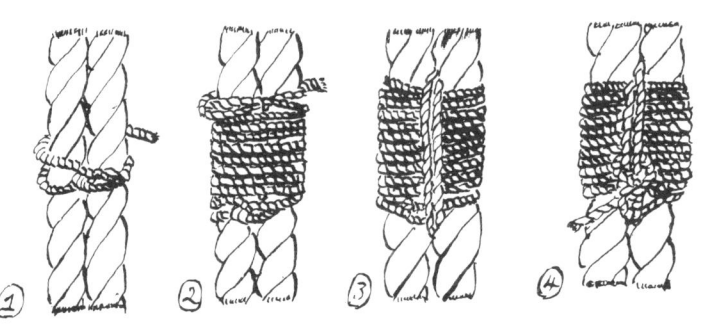

FLAT SEIZING. 1. Make a loop and pass the end through to start. 2. Seize the ropes together. For a ROUND SEIZING come back over the top with riding turns between the first turns. 3. Bind the seizing with cross turns between the ropes. 4. Finish with a clove hitch around the cross turns (frapping).

CROSS SEIZING

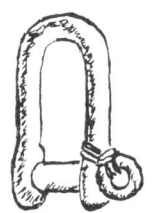

Mousing a shackle with seizing wire.

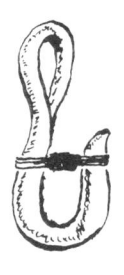

Mousing a hook with marline or spunyarn.

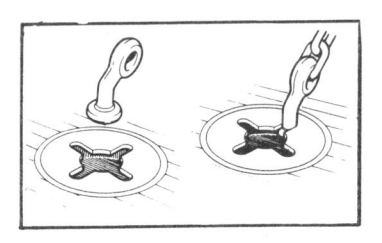

Deck fitting and Elephant's foot sometimes used as an alternative to a ring bolt, as an anchorage for deck cargo lashings.

Plain or common whipping.

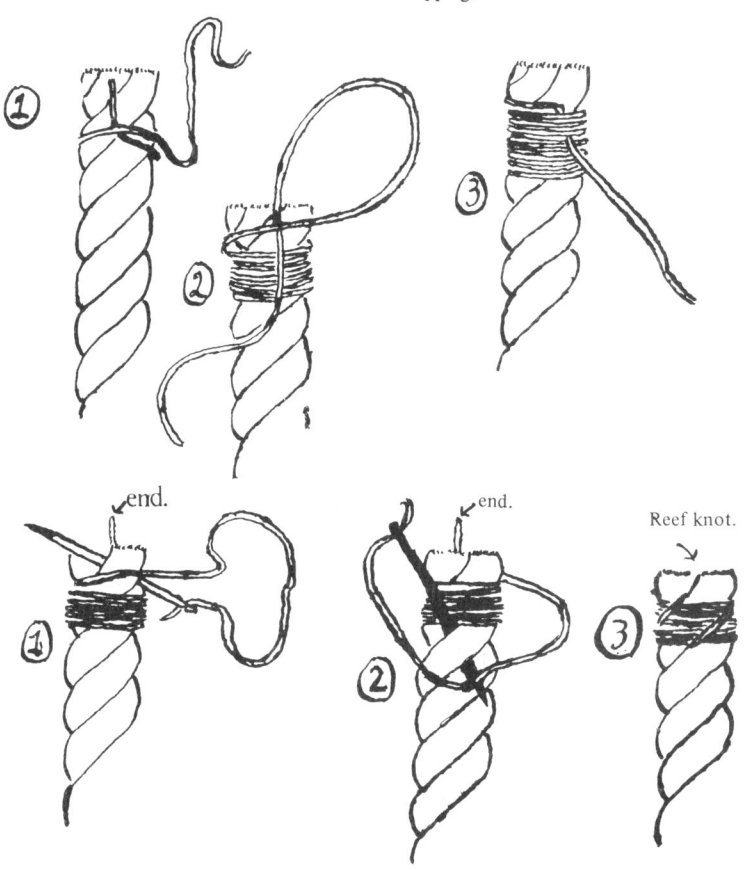

Palm and needle whipping.

(commence the same as a common whipping.)

Racking

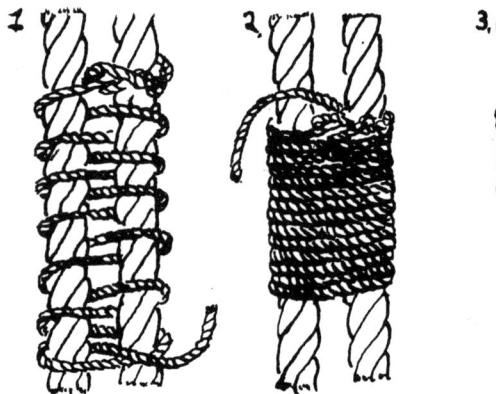

1. Commence by figure of eighting for 8 turns.
2. Complete with a seizing of round turns between the figure of eight turns.
3. Frap and secure with a clove hitch.

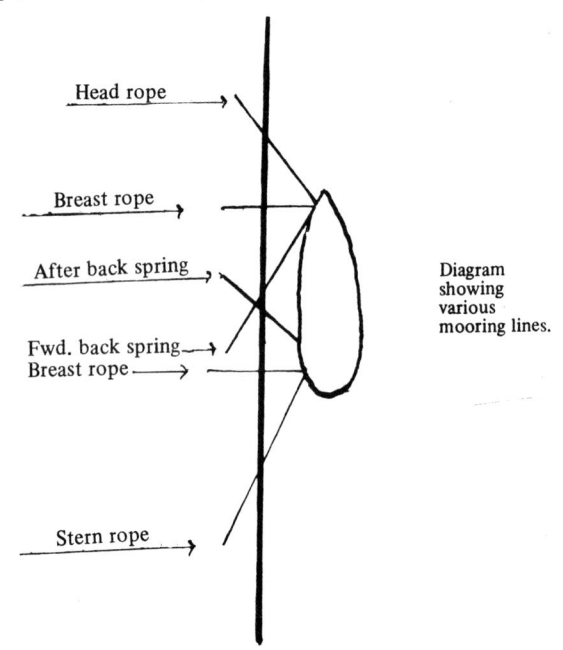

Head rope

Breast rope

After back spring

Fwd. back spring
Breast rope

Stern rope

Diagram showing various mooring lines.

Marrying. To marry the end of a dummy gantline to a gantline for the purpose of reeving the gantline through a sheave. Whip and sew the ends of the two gantlines together end on, with the aid of a palm and needle.

WHIPPINGS.
Plain or common whipping. Temporarily used to prevent the end of a rope from fraying.

Palm and needle whipping. Used to permanently prevent the end of a rope from fraying. Finish off by tying a reef knot between the strands, using both ends of the seaming twine.

SEIZINGS.
Flat seizing. Used to secure two parts of rope together side by side, when the strain on both ropes is equal and in the same direction. Finish off by taking several frapping turns around the seizing, between the ropes and secure the end with two half hitches.

Round seizing. A more secure version of the flat seizing. A flat seizing is made of about 12 turns and riding turns are taken on top and made to lie between the first turns. Several frapping turns are then taken around the seizing between the ropes. Finish off with two half hitches.

Cross seizing. Used to secure the eye of a rope to a particular place on a rope rove through the eye. i.e., for securing the eye of a lifeline to a davit head span.

Racking. Used to secure two parts of rope together side by side, when the strain on the ropes is unequal or in opposite directions.

Mousing. Used to prevent a shackle pin from working out, or to close the eye of a hook so that nothing can become unhooked from it.

Seizings on fibre ropes and hooks are to be made with spunyarn or marline. Seizings on wire ropes and shackles are to be made with seizing wire.

STOPPERS.
Rope stopper. Made of manila or sisal rope it is for use on natural fibre ropes. Not to be used on synthetic fibre ropes.

Chinese or West Country stopper. Made of synthetic or natural fibre rope. Always to be used on synthetic fibre ropes, can be used on natural fibre ropes.

Note that a stopper MUST always be of the same material as the rope it is used to stopper off. However stoppers should not be made of polyamide (nylon) rope. Use polypropylene stoppers on polyamide (nylon) mooring ropes.

Chain stopper. Small chain used to stopper off derrick topping lifts, wire mooring ropes and other wire ropes. The half hitches are separated and the tail is backed against the lay of the wire to ensure that the chain neither jams or opens up the lay of the wire.

Carpenter's stopper. An efficient mechanical means of stoppering off any wire rope. Made in various sizes.

CARPENTER'S STOPPER FOR USE ON WIRE ROPE O[N]

PIN
WEDGE
KEEP PLATE
HOLD HERE
HINGE

ROPE STOPPER FOR NATURAL FIBRE ROPE ONLY
HALF HITCH

CHINESE OR WEST COUNTRY STOPPER FOR SYNTHETIC FIBRE ROPE OR NATURAL FIBRE ROPE
Twist and hold
underneath rope twisted with the lay.
Top rope twisted against the lay
HALF HITCH

CHAIN STOPPER FOR WIRE ROPE ONLY
Cord rope tail.
Cow hitch

The correct method of applying "Bull Dog" grips: at least three — more if required.
'U'-bolts on short end. — Castings on standing part.

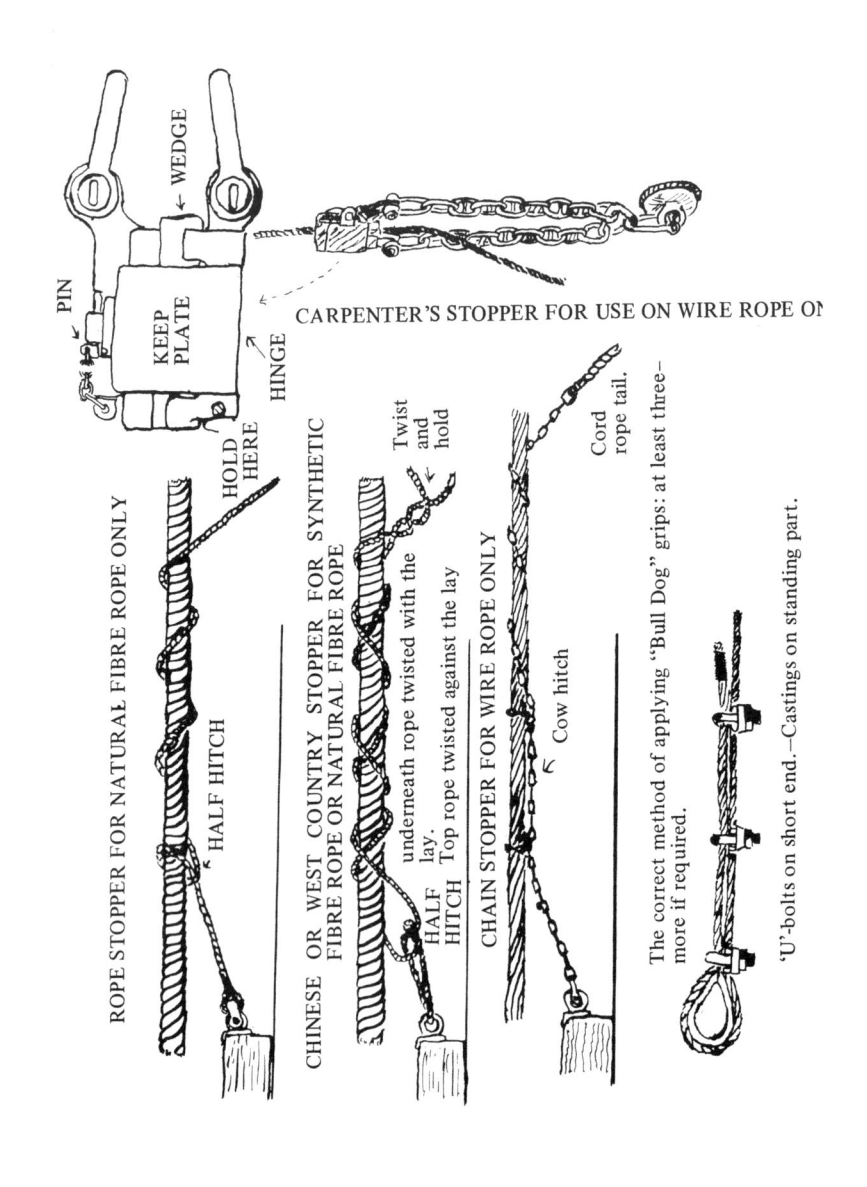

144

APPLICATION

Riding a stay. Travelling the length of a stay, shroud or swifter. Sitting in a bosun's chair that is attached to a shackle through which the stay passes, for the purpose of oiling or painting the stay. The pin of the shackle MUST go through the bridle of the chair and be moused.

Ladders. When using a ladder it must be lashed TOP and BOTTOM. In port when a ladder has to be used in lieu of a gangway. The ladder should if possible be at a near vertical angle, the top is to be lashed by taking turns around the whole ladder and NOT between the rungs. This allows the ladder to slide up or down the bulwarks as the water level or the amount of the ship's freeboard changes. A lifebuoy with a line and a net are to be provided. At night the ladder is to be well lighted. A Bulwark ladder should be secured adjacent to the ladder.

Merchant Shipping Notice No. M.898

PILOT LADDERS AND MECHANICAL PILOT HOISTS

Notice to Owners, Masters, Shipbuilders and Manufacturers

This Notice supersedes Merchant Shipping Notice M.753

1. Attention is drawn to IMCO Resolution A.275 (VIII) (20 November 1973) Recommendation on Performance Standards for Mechanical Pilot Hoists, which sets out performance criteria for the design, construction, testing and operation of pilot hoists. These recommended standards are reproduced as the Annex to this notice.

2. The Department will recommend the acceptance only of new designs of pilot hoists which meet these standards in every respect. Manufacturers of pilot hoists are advised to submit to the Department designs that are in strict accordance with the relevant provisions.

3. It is strongly recommended that hoists manufactured prior to the Resolution should remain in service only if they are modified, if necessary, to incorporate, inter alia, the following features:

   (a) arrangements such that failure of one fall wire or end fixing will not leave the ladder section unsupported;

   (b) a ladder section constructed in accordance with paragraph 2.4 of the resolution;

   (c) efficient arrangements to ensure that the falls wind evenly on to the winch drum;

   (d) a brake or other equally effective arrangement such as a properly constructed worm drive, which is capable of supporting the working load in the event of power failure;

   (e) an emergency stop switch within easy reach of the pilot by means of which he may cut off the power.

4. Where a pilot hoist is provided, personnel engaged in rigging and operating it should be fully instructed in the safe procedures to be adopted and the equipment should be tested prior to use. The length of ladder referred to at

paragraph 2.5(c) of the resolution should be such that access to it is available from the pilot hoist during any point of its travel.

5. In addition to the testing required by paragraph 2.6 of the resolution, regular test rigging and inspection, including a load test to at least 150 kgs should be carried out by the ship's personnel at intervals of not more than six months, and a record to that effect maintained by the master, in the ship's log.

6. Provision of a pilot hoist does not relieve owners and masters of the statutory duty to provide embarkation and disembarkation arrangements complying with the requirements of the Merchant Shipping (Pilot Ladders) Rules 1965 as amended.

7. Further, the Department wishes to stress the need for strict compliance with the Pilot Ladder Rules in order to minimise the danger to pilots when boarding and leaving ships. Particular attention should be given to the following points:—
   (a) Pilot ladders should be rigged in such a manner that the steps are horizontal, and such that the lower end is at a height above the water to allow ease of access to and from the attendant craft;
   (b) Pilot ladders constructed with large wooden inserts above and below the steps cause difficulty in grasping the side ropes and their use is not recommended. Small winnets however, are acceptable;
   (c) When an accommodation ladder is used in conjunction with a pilot ladder, it is essential that the pilot ladder should be positioned in such a manner as to afford easy and safe access to the accommodation ladder platform;
   (d) The rigging of pilot ladders and the embarkation and disembarkation of pilots must be supervised by a responsible officer of the ship.

8. The offering of a proper lee to the pilot boat by the ship is of great importance. The arrangements for boarding should preferably be sited as near amidships as possible, but in no circumstances should they be in a position which could lead the pilot boat to run the risk of passing underneath overhanging parts up the ship's hull structure.

ANNEX

Recommendation on Performance Standards for Mechanical Pilot Hoists
(IMCO Resolution A.275 (VIII) Adopted on 20 November 1973)

1. **General**
1.1 Mechanical pilot hoists and ancillary equipment should be of such design and construction as to ensure that the pilot can be embarked and disembarked in a safe manner. The hoist should be used solely for the embarkation and disembarkation of personnel.
1.2 The working load should be the sum of the weight of the ladder and falls in the fully lowered condition and the maximum number of persons which the hoist is designed to carry, the weight of each person being taken as 150 Kgs.
1.3 Every pilot hoist should be of such construction that when operating under the defined working load each component should have an adequate factor of safety having regard to the material used, the method of construction and the nature of its duty.
1.4 In selecting the materials of construction, due regard should be paid to the conditions under which the hoist will be required to operate.
1.5 The pilot hoist should be located within the parallel body length of the ship and clear of all discharges.
1.6 The operator should be able to control the hoist when he is in a standing position and looking over the ship's side for observing the hoist, even in its lowest position.

1.7 The manfacturer of the pilot hoist should supply with each installation an approved maintenance manual, together with a maintenance log;

Each installation should be kept in good order and maintained in accordance with the instructions of the manual. All maintenance and repairs of the installation should be recorded in the log.

2. Construction

2.1 The hoist should generally consist of the following three main parts, but hoists of other equally effficient construction may be considered:
  (a) a mechanical powered appliance together with means for a safe passage from the hoist to the deck and *vice versa*;
  (b) two separate falls;
  (c) a ladder consisting of two parts;
    (i) a rigid upper part for the transportation of the pilot upwards or downwards;
    (ii) a lower part consisting of a short length of pilot ladder, which enables the pilot to climb from the pilot launch to the upper part of the hoist and *vice versa*.

2.2 Mechanical powered appliance

(a) The source of power for the winches may be electrical, hydraulic or pneumatic. In the case of an pneumatic system an exclusive air supply should be provided with arrangements to control its quality. It may be necessary to give special consideration to the selection of the type of source of power for ships engaged in the carriage of flammable cargoes. All systems should be capable of efficient operation under the conditions of vibration, humidity and change of temperature likely to be experienced in the vessel in which they are installed.

(b) The design of the winch should include a brake or other equally effective arrangement such as a properly constructed worm drive, which is capable of supporting the working load in the event of power failure.

(c) Efficient hand gear should be provided to lower or recover the pilot(s) at a reasonable speed in the event of power failure. The brake or other arrangement in sub-paragraph (b) above should be capable of supporting the working load when the hand gear is in use.

(d) Crank handle(s) provided for manual operation should, when engaged, be interlocked so that the power supply is automatically cut off.

(e) Hoists should be fitted with safety devices to automatically cut off the power supply when the ladder comes up against any stop to avoid overstressing the falls or other parts of the hoist. However, in the case of hoists operated by pneumatic power, if the maximum torque available from the air motor cannot result in overstressing of the falls or other parts of the hoist, the safety cut-out device may be omitted.

(f) All hoist controls should incorporate an emergency stop to cut off the power supply.

(g) The winch controls should be clearly and durably marked to indicate the action to "Hoist", "Stop", and "Lower". The movement of these controls should correspond with the movement of the hoist returning to the stop-position when released.

(h) Efficient arrangements should be provided to ensure that the falls wind evenly on to the winch-drums.

(i) Pilot hoists should be securely attached to the structure of the ship. Proper and strong attachment points should be provided for hoists of the portable type on each side of the ship. Attachment of the pilot hoist should not be solely by means of the ship's side rails.

(j) The winch should be capable of hoisting or lowering the pilot(s) at a speed of between 15 and 30 metres per minute.

(k) There should be safe means of access between the top of the hoist and the deck and *vice versa*; such access should be gained directly by a platform securely guarded by handrails.

(l) Any electrical appliance associated with the ladder section of the hoist should be operated at a voltage not exceeding 25 volts.

2.3 Falls

(a) Two separate wire rope falls should be used, made of flexible steel of adequate strength and resistant to corrosion in a salt-laden atmosphere.

(b) Wire ropes should be securely attached to the winch-drums and the ladder. These attachments should be capable of withstanding a proof load of not less than 2·2 times

the load on such attachments. The falls should be maintained at a sufficient relative distance from one another.
(c) The wire rope falls should be of sufficient length to allow for all conditions of freeboard encountered in service and to retain at least three turns on the winch-drums with the hoist in its lowest position.

2.4 Ladder section

The laddr section should comprise a rigid and a flexible part, complying with the following requirements:
(a) The rigid part should be not less than 2·50 metres in length and be equipped in such a way that the pilot can maintain a safe position whilst being hoisted or lowered. Such parts should be provided with:
   (i) a sufficient number of steps to provide a safe and easy access to and from the platform referred to in paragraph 2.2, sub-paragraph (k);
   (ii) suitable protection against extremes of temperature to provide safe handholds and fitted with non-skid steps;
   (iii) a spreader at the lower end of not less than 1·80 metres. The ends of the spreader should be provided with rollers of adequate size which should roll freely on the ship's side during the whole operation of embarking or disembarking;
   (iv) an effective guard ring, suitably padded, so positioned as to provide physical support for the pilot without hampering his movements;
   (v) adequate means for communication between the pilot and the operator and/or the responsible officer who supervises the embarkation or disembarkation of the pilot;
   (vi) whenever possible an emergency stop switch within easy reach of the pilot by means of which he may cut off the power.
(b) Below the rigid part mentioned in sub-paragraph (a) above, a section of pilot ladder comprising 8 steps should be provided, constructed in accordance with the following requirements:
   (i) The steps of the pilot ladder should be:
      (1) of hardwood, or other material of equivalent properties, made in one piece free of knots, having an efficient non-slip surface; the four lowest steps may be made of rubber of sufficient strength and stiffness or of other suitable material of equivalent characteristics.
      (2) not less than 480 millimetres long, 115 millimetres wide, and 25 millimetres in depth, excluding any non-slip device;
      (3) equally spaced not less than 300 millimetres nor more than 380 millimetres apart and be secured in such a manner that they will remain horizontal.
   (ii) No pilot ladder should have more than two replacement steps which are secured in position by a method different from that used in the original construction of the ladder and any steps so secured should be replaced as soon as reasonably practicable by steps secured in position by the method used in the original construction of the ladder. When any replacement step is secured to the side ropes of the ladder by means of grooves in the sides of the step, such grooves should be in the longer sides of the step.
   (iii) The side ropes of the ladder should consist of two uncovered manilla ropes not less than 60 millimetres in circumference on each side. Each rope should be continuous with no joins below the top step.
(c) The top step of the flexible pilot ladder and those of the rigid ladder should be in the same vertical line, of the same width, spaced vertically equidistant and placed as close as practicable to the ship's side. The handholds of both parts of the ladder should be aligned as closely as possible.

2.5 Operational aspects

(a) Rigging and testing of the hoist and the embarkation and disembarkation of a pilot should be supervised by a responsible officer of the ship. Personnel engaged in rigging and operating the hoist should be instructed in the same procedures to be adopted and the equipment should be tested prior to the embarkation or disembarkation of a pilot.
(b) Lighting should be provided at night such that the pilot hoist overside, its controls and the position where the pilot boards the ship should be adequately lit. A lifebuoy equipped with a self-igniting light should be kept at hand ready for use. A heaving line should be kept at hand ready for use if required.

(c) A pilot ladder complying with the provisions of Regulation 17, Chapter V, of the International Convention for the Safety of Life at Sea 1960, should be rigged on deck adjacent to the hoist and available for immediate use.
(d) The position on the ship's side where the hoist will be lowered should be indicated as well as possible.
(e) An adequate protected stowage position should be provided for the portable hoist. In very cold weather to avoid the danger of ice formation, the portable hoist should not be rigged until use is imminent.
(f) The assembly and operation of the pilot hoist should form part of the ship's routine drills.

2.6 **Testing**
(a) Every new pilot hoist should be subjected to an overload test of 2·2 times the working load. During this test the load should be lowered a distance of not less than 5 metres.
(b) A operating test of of 10 per cent overload should be carried out after installation on board the ship to check the attachment and performance of the hoist to the satisfaction of the Administration.
(c) Subsequent examinations of the hoists under working conditions should be made at each survey for the renewal of the vessel's safety equipment certificate.

STAGES

When painting overside on a stage. Never rig a stage over a dry dock, quay, barge or lighter alongside. Use a stage only when there is water beneath.

Paint pots (not too full) should be let down to the required height by their own lanyards, from the upper deck and be kept well clear of the stage gantlines, in order to try and ensure that the gantlines do not become contaminated with wet paint. All brushes and other equipment should be attached to lanyards for the purpose of securing the equipment when not in use. One lowering hitch should be led in-board and the other outboard, to avoid tipping the stage.

A side ladder is to be rigged conveniently either just forward or aft of the stage to give suitable access safe to the stage.

At least two gantlines each long enough to trail in the water, should be rigged at the position of the stage, both for operators on the stage to hold in order to help maintain their balance and to catch hold of, if either by design or accident, a man falls in the water. This especially necessary when either, a non-swimmer is on the stage or a current is flowing past the ship.

When painting over the bow and under the quarter, bowsing-in lines, made fast to the stage ends and taken forward and aft, will be needed to bring the stage close to the ship's side. These should be of the same strength as the gantlines. A back rope (again of equal strength) should also be rigged between the gantlines to give the operators additional support.

When using either stages or bosun's chairs to paint the funnel, safety belts must be worn properly adjusted, so as to prevent a man falling to the deck in the event of an accident of any description. Do not rely on blocks permanently fixed to the funnel top. Use proper **portable funnel blocks** and unship them again after use. Blocks remaining at the funnel top are prone to corrosion from chemical fumes, heat, and rust, with the result that sometimes they are far from trustworthy. The engine room is to be warned before the job is started and again when finished.

STAGE KNOT WITH LOWERING HITCH.

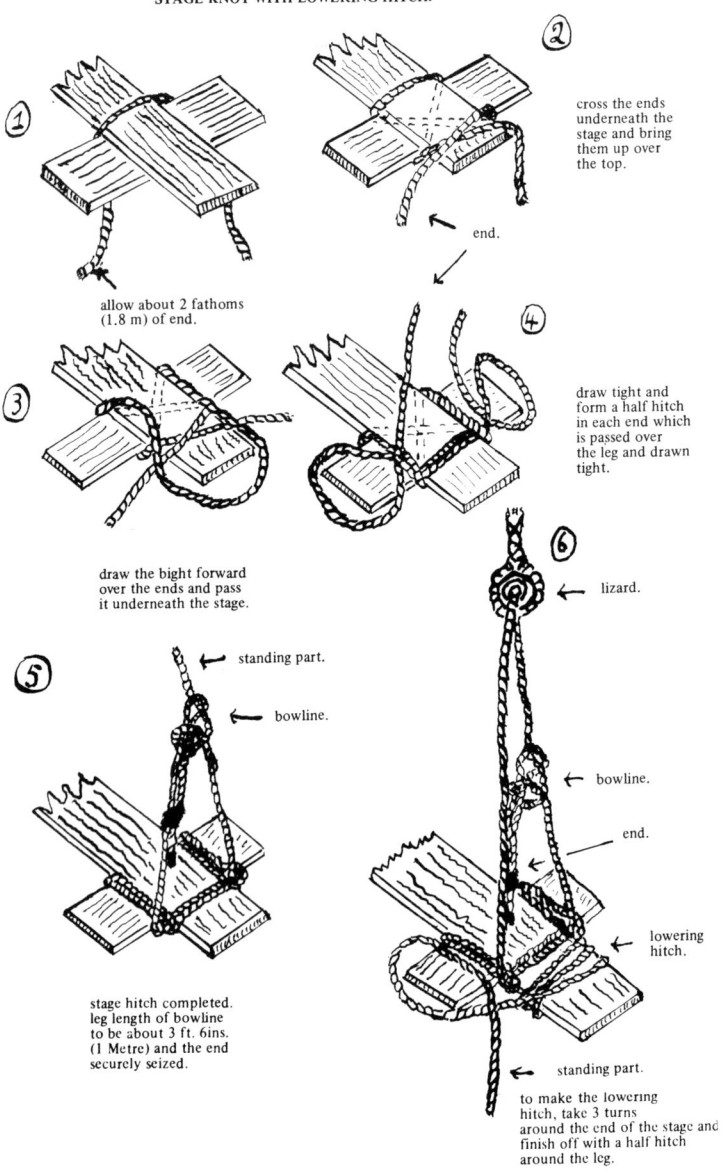

①

② cross the ends underneath the stage and bring them up over the top.

← end.

allow about 2 fathoms (1.8 m) of end.

③

④ draw tight and form a half hitch in each end which is passed over the leg and drawn tight.

draw the bight forward over the ends and pass it underneath the stage.

⑤

← standing part.

← bowline.

stage hitch completed. leg length of bowline to be about 3 ft. 6ins. (1 Metre) and the end securely seized.

⑥ ← lizard.

← bowline.

← end.

← lowering hitch.

← standing part.

to make the lowering hitch, take 3 turns around the end of the stage and finish off with a half hitch around the leg.

Accommodation Ladder
To rig an accommodation ladder.

1. Take a 3" (24 mm) fibre rope and secure the end to a strong point inside the bulwarks. Pass the bight of the rope around the end of the accommodation ladder.
2. Secure the standing part of the rope to a strong cleat, so that the rope will take the weight when the clips or lashings on the accommodation ladder are let go.
3. Rig a handy billy between a ringbolt and the platform.
4. Let go all other lashings, clips and bolts.
5. Turn out the platform and accommodation ladder.
6. Make the bridle fast to the gantry and turn it out.
7. PUT ON A SAFETY BELT OR LIFELINE. INFORM THE BRIDGE THAT A MAN IS GOING OVERSIDE.
8. On the accommodation ladder make fast the bridle shackles, ship the stanchions and set up the man ropes.
9. Lower the accommodation ladder by means of the rope, until the weight is taken on bridle.
10. Remove the rope.
11. Set up a safety net beneath the accommodation ladder and place a lifebuoy with a line attached, handy to the platform. Rig suitable lighting.
N.B. When light ship, it may be necessary to rig an extension on to the bottom of the accommodation ladder.

Awnings.
When taking in a canvas awning, let go the lee side and gather it in to windward. Similarly when stretching, make the weather side fast first and stretch to leeward.

Moorings.
Entering port, in preparation take to the fo'castle head and poop, two heaving lines, ratguards, two cork fenders, two suitable rope stoppers, two suitable chain stoppers, an oil can and/or grease gun and a snatch block for the back springs if a suitable lead is not provided. Note that if a monkey's fist, hangman's knot or other fancy knot is made in the end of a heaving line to help carry the line, it should not be weighted.

Unlash the mooring lines and take the tarpaulins off, or bring them from their stowed position and coil them clockwise on duck boards in suitable positions. Take the covers off the wire reels (stow the covers in the bo'sun's store) and take at least one wire off its reel, flaking the wire up and down the deck (clear of space required for working), ready to run out. Request power on deck from the engine room, oil all roller fairleads and the working parts of the windlass, capstan, or winch. See that all working space is clear.

The Welin Automatic System for Fixed Length Accommodation Ladders

SHOWING THE CORRECT USE OF SLINGS AND STROPS.
VARIOUS ANGLES.

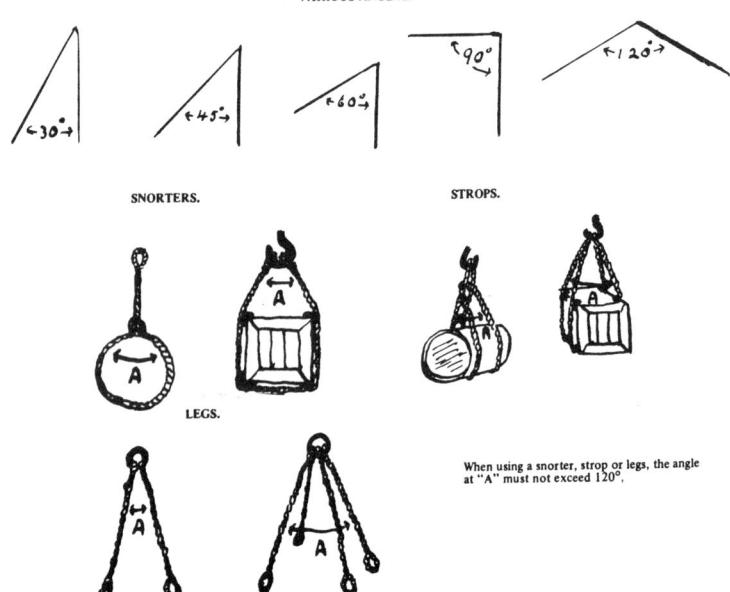

SNORTERS.

STROPS.

LEGS.

When using a snorter, strop or legs, the angle at "A" must not exceed 120°.

The above sketches show the correct use of snorters and strops. All sharp angles on loads must be packed in order to ensure that (a) the strop does not cut the packaging and (b) That sharp corners do not cut the snorter or strop. Wide canvas slings are to be used on loads that are liable to damage i.e., cement in paper bags etc.

OIL JETTY GANGWAY SYSTEM

To join two mooring lines. Seize the eyes together using a rope stopper.

To send a mooring line ashore. Pass a heaving line through the eye and secure with a bowline. Throw the other end of the heaving line ashore.

MAKING FAST TO THE BITTS.

With a synthetic fibre rope always take two round turns around the leading bitt, before figure of eighting.

When making fast a wire rope to the bitts always lash the top turns down, to prevent the wire springing off the bitts.

When placing the eye of a second mooring line over a bollard. Take the eye of the second line up through the eye of the first line.

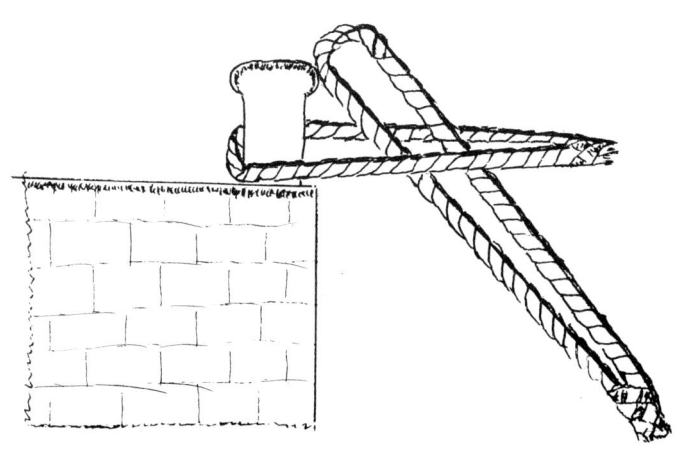

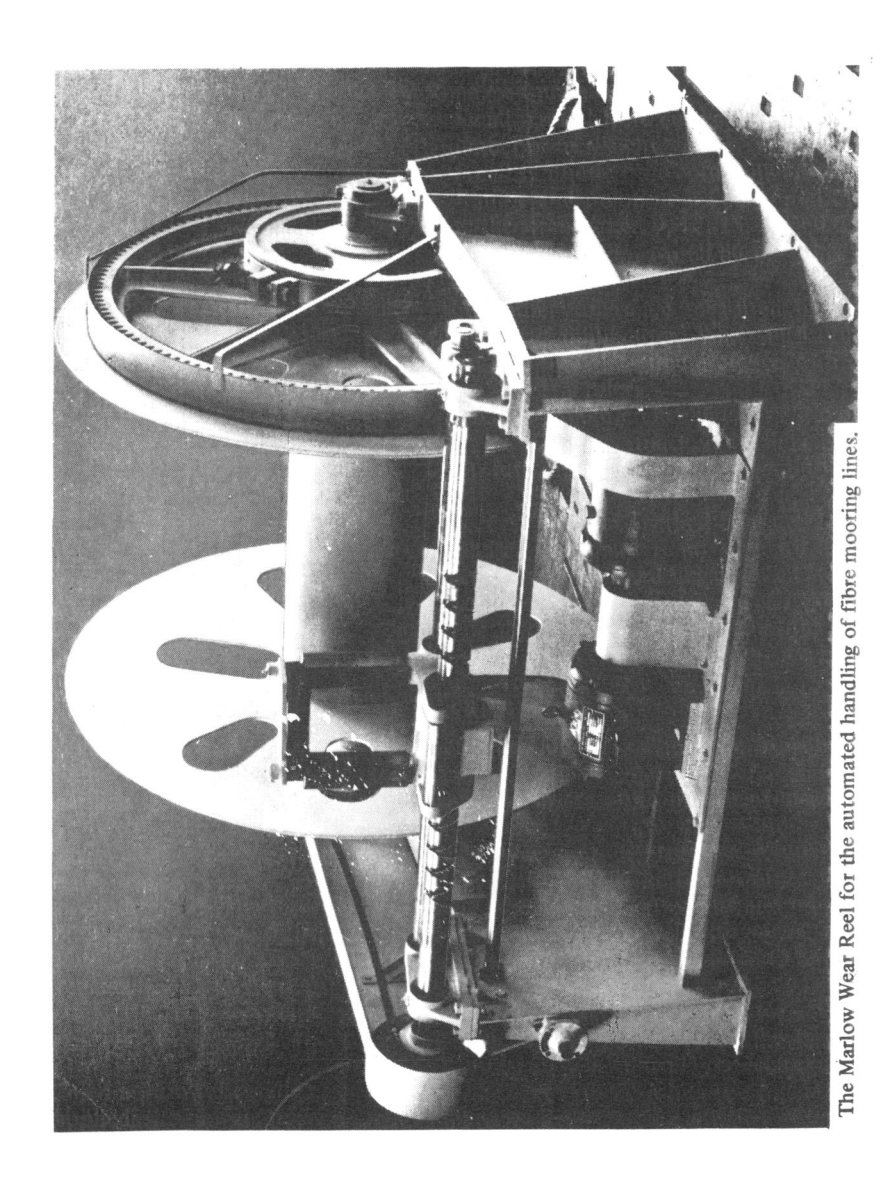

The Marlow Wear Reel for the automated handling of fibre mooring lines.

Putting out a fibre towing spring with a wire rope tail, it is advisable to have a chain stopper loosely placed and ready around the wire rope. As the tug goes away the weight of the tail is more than can be held by hand, catch the wire in the stopper and it will hold the wire while it is being made fast to the bitts, let go the stopper as soon as the wire is fast. Some ships have a rope messenger the length of the towing spring spliced into the towing eye of the rope. When the tug lets go, haul in by taking the messenger to the drum-end. With the rope aboard, it is an easy matter to bring the wire aboard and a lot quicker aft, where the 2nd. Mate is in a hurry to get the towing spring aboard and clear of the propeller.

When warping the ship alongside, have a man on the windlass or winch and another man to stow the rope as it comes in, keep out of any bights. With synthetic fibre ropes never take more than three turns on the drum-end or capstan unless it is whelped. Do not surge the rope especially when it has weight on it, stop the winch and walk back if necessary. Beware of synthetic rope mooring line when it has a very heavy strain on it. They sometimes jump round the drum-end to regain elasticity and always take a fair amount of rope very quickly, when they do. Stand well back from the drum-end so that your hand cannot get caught between rope and drum-end.

Use manila rope stoppers on manila lines and polypropylene chinese stoppers on polyamide (nylon) or polypropylene lines, chain stoppers on wire ropes. (Never make a rope stopper fast with a rolling hitch, this is very bad practice).

When making fast a synthetic rope, always take at least two round turns around the nearest bitt, before figure-of-eighting. When a wire rope has been made fast to the bitts, always lash the top turns down to prevent them springing off the bitts. While figure-of-eighting a heavy wire rope, one man should hold each turn down as it is put on the bitts by another man. Wire ropes have a nasty habit of springing off the bitts and deliberately hitting you smack in the face, hard.

Never take any rope to a drum-end without a good lead. If a suitably placed roller lead is not provided, use a snatch block strategically placed. Be sure that the snatch block, shackle and ring bolt are all stronger than the rope.

Wear suitable footwear, and gloves when handling a wire rope or in arctic conditions. Gloves are not advised for use when handling fibre ropes unless weather conditions make it essential.

When putting out a bight and when putting a slip wire onto a buoy, seize the two parts of the eye together and make the bare end of the rope fast to the bitts by figure-of-eighting. **Do not put the eye over the bitts.**

When the ship is tied up, put rat guards on all ropes and wires, if the ship is alongside a quay or jetty. Reel up the wires and coil down fibre ropes neatly, cover fibre ropes with tarpaulins. Return all the gear and notify the engine room "finished with power".

Leaving port. take the same gear forward and aft, remove the tarpaulins from the ropes and take off the rat guards. Obtain power on deck and oil all working parts.

Keep your working space clear all the time, coil rope and reel up wire as it comes in. Allow the ends of all ropes and wires to come round the drum-end, this helps to keep kinks out of them. Finally lash and cover all fibre ropes or stow them away. Reel up all wires and replace the reel covers in position. Return all gear and notify the engine room "finished with power".

Warping. Whenever the ship is being warped along the quay and most especially if this is done as soon as the ship comes alongside, in order to get the ship correctly placed alongside the shore equipment. No mooring rope, wire or spring is to be used to check the ship while it is on a winch-drum end. Take the rope, wire or spring off the winch drum-end and catch a few turns around the bitts with it, before using it as a check rope. In the case of a wire rope, ensure that ample wire has been taken off the reel and that it is clear for running out.

ACCIDENT PREVENTION.

Always examine and test a gantline before using it to go aloft.
Always when wearing gloves, see that they are loose fitting and can easily pull off.
Always wear gloves when handling wire rope.
Always keep all ropes neatly coiled and clear of working spaces.
Always remove more than sufficient wire rope from the reel before using a wire rope.
Always have a good sharp sheath knife on your belt. (A cobbler's knife in a home made sheath is as good as any) Dull knives are useless in a sudden emergency.
Never stand in the bight of a rope.
Never stand near or in line with a rope under strain.
Never surge or render a synthetic rope.–It will fuse.–Stop the winch and if necessary, 'walk back the winch'.
Never handle a rope on a winch drum end or barrel, without someone in control of the winch.
Never make any rope fast to a winch drum end.
Never make a stage gantline, side ladder, guy fall, lizard or any other rope fast to a portable ship's side rail.
Never take more than three turns of a synthetic mooring rope on a winch drum end when warping, unless the drum is whelped.
Never use a snatch block that has not been tested to a greater breaking strain than the rope you intend to use and be sure that the ringbolt and shackle are equally strong.
Never allow wires to cross fibre ropes on bitts, bollards or fairleads.
Remember that a synthetic rope gives no visible or audible warning before parting.
Remember that the beds of the oceans are littered with seamen, whose famous last words have been "I know what I'm doing." Always wear protective clothing when required and a safety belt or harness whenever you are aloft, overside or working in a peak or tank.
Above all, do not have loose clothing that can fall into machinery, or wear unsuitable footwear, or rings.
When using a lizard with a lowering hitch on a stage. If the lizard is led over a sharp plate edge, a channel, or some other means must be used to ensure that the lizard does not cut or fray where it passes over the plate edge.
Whenever you are working on a stage and not wearing a safety harness - the stage should be properly fenced to a height of 3 feet (90 cms.)

Merchant Shipping Notice No. M.544.

WEARING OF SAFETY BELTS OR HARNESSES.

In a recent fatal accident a member of the crew of a ship was swept away after falling from an accommodation ladder which was being lowered while the ship was under way.

The Board are aware that in exceptional circumstances, it may be necessary for men to work overside whilst a vessel is under way and, in order to reduce the risk of similar accidents, the wearing of a safety belt or harness attached to a lifeline is recommended.

Moreover, whenever it is necessary to send a man overboard whilst the ship is under way, the officer of the watch should be notified and assured that the safety belt or harness attached to a lifeline is being worn and is being properly tended.

A safety belt or harness complying with BS 1937: 1967 (Type 3–general purpose safety belts and harnesses incorporating flexible connections for attachment to anchorage points) could be a suitable appliance to use for general purpose work on board ship. However, where such an appliance is so provided, it should be stowed in a place where there is little chance of it being confused with a similar appliance supplied to ships in compliance with the Merchant Shipping (Fire Appliances) Rules 1965 (SI 1965/1106).

October 1968.

Merchant Shipping Notice No. M.547.

ACCIDENTS INVOLVING USE OF MOORING WIRES AND REELS.

Accidents to seamen occur frequently both as a result of using a mooring wire directly from a reel or coiling it on deck adjacent to a cleat or bitts from which it is being rendered, either as a check wire or spring or in lowering some heavy object.

Using a wire direct from the reel has been known to result in the reel and frame being torn from the deck when the wire fouled the reel causing considerable damage as well as injury and death to seamen on deck. In a more recent case of the reel being fouled by the wire and no slack being available, a small towing vessel was made to heel over, to become swamped and as a result, to sink, drowning the crew.

The practice of coiling a wire close to the cleat or bitts from which it is being rendered, can lead to serious injury or even death to men in consequence of their being caught in the loose coils, should the wire suddenly become unmanageable and run out of control.

It must therefore be emphasised that both of these practices are unseamanlike and extremely dangerous.

In no circumstances should a wire be used direct from a reel to a set of bitts or cleats when the wire has to be rendered under tension. Sufficient wire should be taken from the reel for the purpose intended (if necessary removing all the wire from the reel) and flaked on deck well clear of the working area. If possible a second seaman should be available to feed the slack to the seaman rendering the wire round the bitts or cleat. Before the weight is taken, the second seaman should ensure that there are sufficient turns on the bitts or cleat for the purpose without danger of jamming.

January 1969.

Merchant Shipping Notice No. M.718

MOORING, TOWING, HAULING EQUIPMENT ON ALL SHIPS

Notice to Builders, Owners, Masters or Skippers, Officers and Men of Merchant Ships and Fishing Vessels

1. Operations such as mooring, towing and trawling impose very great loads upon ropes or warps, gear and equipment. The Code of Safe Working Practices for Merchant Seamen and the Code of Safety for Fishermen set out certain precautions which should be taken but the circumstances of recent accidents show that greater emphasis should be given to considering the system as a whole.

2. Because of the imposed loads, sudden failure in any part of the system may cause death or serious injury to personnel. Preferably winches or windlasses should be constructed to give warning of undue strains by stalling at well below half the designed maximum safe working load of the weakest element in the system and to afford further protection by walking-back at about half the design safe load. Where that is impracticable, the layout of the installations should be such as to avoid men being stationed or necessarily working in the bight or warp or rope formed by the lead from the winch or windlass round and through the fairleads and over-side. In any case, the consequences of failure in any part of the system must be carefully considered and effective precautions taken.

3. Particular attention is drawn to the need to ensure that pedestal roller fairleads, lead bollards, mooring bitts etc. are (*a*) properly designed to meet all forseeable operational loads and conditions, (*b*) correctly sited and (*c*) effectively secured to a part of the ship's structure which is suitably strengthened. Investigation of one accident showed that due to corrosion fatigue a roller pin fractured at a sharp change of section machined at the lower end. The place of fracture was inacccessible to inspection and maintenance being just below the housing surface. In another instance, the welding between fairlead pedestal and deck failed. It is essential that such welding should be preceded by careful preparation of the plate edges and carried out by a fully competent welder. In a third case, a bollard which was pulled out had been secured to a deck pad by bolts of inadequate diameter and loose nuts.

4. All fixed and running gear including ropes should be *carefully maintained* and *regularly inspected* against wear, damage and corrosion. At all times when the gear is under load, men essential to the operation should be as far as possible in a protected position and others should keep clear of the area. Immediate action should be taken to reduce the load should signs of excessive strain appear in any part of the system.

Merchant Shipping Notice No. M.605.

CARE AND INSPECTION OF GANTLINES USED WITH BOSUN'S CHAIRS, SAFETY HARNESS LINES AND STAGE ROPES.

Seamen have been killed through the sudden parting of the ropes suspending the stages or bosun's chairs from which they were working. Subsequent investigations have shown that the ropes had been impaired by contamination, but that the resultant damage had been practically undetectable by prior visual examination. In three recent cases the rope that parted was 2" or 2½" sisal which had been contaminated with phosphoric acid, an acid present in most rust-removing preparations.

1. Ropes, whether of man-made or natural fibre, should be stowed separately from any container of acids, alkalis, detergents or other cleaning fluids and rust-removing preparations, and protected from possible contamination by these substances.
2. Man-made fibre ropes, which should be stowed away from sources of heat and strong sunlight, should also be kept from contact with wet paint, varnish, coal tar, paint thinners or paint stripping preparations. In any operation involving the use of these contaminating substances, care should be taken to avoid ropes being splashed or wetted by them. Should a rope become contaminated it should not be further used for such purposes, unless the contamination is superficial, when the rope should be thoroughly washed with water as soon as possible.
3. Ropes used as gantlines to support Bonun's chairs, stage ropes, safety harness lines, and lifelines should be load tested to four or five times the loads they will be required to carry just prior to use if there is the slightest possibility that they have been contaminated.
4. Any blocks and lizards which are used should be carefully examined before use.

December 1970.

BACK SPLICE

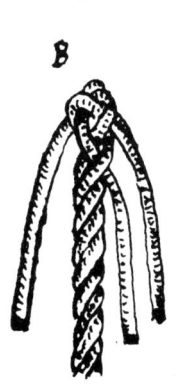

A. Make a crown. B.
B. With the crown towards you, tuck each end over one and under one.
C. Repeat, making two tucks in all.

CHAPTER 9.
SPLICING.

Back splice.	The end of a fibre rope is tucked back into itself, to prevent the rope unlaying.
Cut splice.	A form of short splice, each end of the rope is spliced into the other rope a little distance from the end, thus leaving a small length of rope where the two ropes lie alongside each other. In other words tuck the tails of each rope into the other, as though you were making two eye splices. (Useful for wire and plaited ropes).
Eye splice.	A loop or eye is formed in the end of a rope. The end is tucked back into the rope to make the eye permanent. Note:— That whenever a wire rope is spliced into an eye at the end of a fibre rope, the fibre rope eye must contain a thimble. Left hand laid ropes must never be joined to right hand laid ropes.
Long splice.	Made by laying up the strands from each rope into the other rope by replacing one strand with another. It is necessary where the rope will have to pass over a sheave. Very wasteful of rope and seldom used at sea.
Short splice.	Two ends of rope are interlaced to make a continuous rope.

TO SPLICE ANY HAWSER LAID ROPE.

First get your tools together, you will need a sharp knife, some seaming twine and for ropes of 2½ inch circumference (20mm) and over a suitable fid, add a wood mallet for mooring lines.

A fid as used for rope work is a smooth conical piece of hard wood, usually lignum-vitae, the base should be flat. Made in various sizes, it is used to pick up and separate the strands of a rope, one at a time. A ferule fid (used when splicing synthetic fibre ropes) consists of an ordinary fid that has a length of hollow tubing extending from its base. The fid is used to pick up a strand and open the rope. When splicing synthetic fibre rope, the tail to be tucked is then pushed into the open end of the hollow tube at the base of the fid and drawn through the rope with the fid (this is necessary because synthetic fibres are much softer than natural fibres). Never hammer a fid into a rope, in cases of difficulty place the flat base of the fid on the deck and hammer the rope down the fid with a wood mallet.

Measure off six inches (15cm) of rope for every inch (2.5cm) of the rope's circumference (five to six turns) from the end and put a good tight whipping on the rope at this point. Unlay the strands as far as the whipping and whip the tails tightly. On synthetic fibre ropes the tails may be heat sealed but it is advisable with snythetic fibre ropes to put a few spaced whippings or a lacing along the length of each tail, to prevent the yarns, which are very fine, from becoming partly unlaid.

Back splice. Put a tight whipping on each strand and unlay about three turns of rope. With the three tails form a crown and pull it tightly home. Tuck each tail

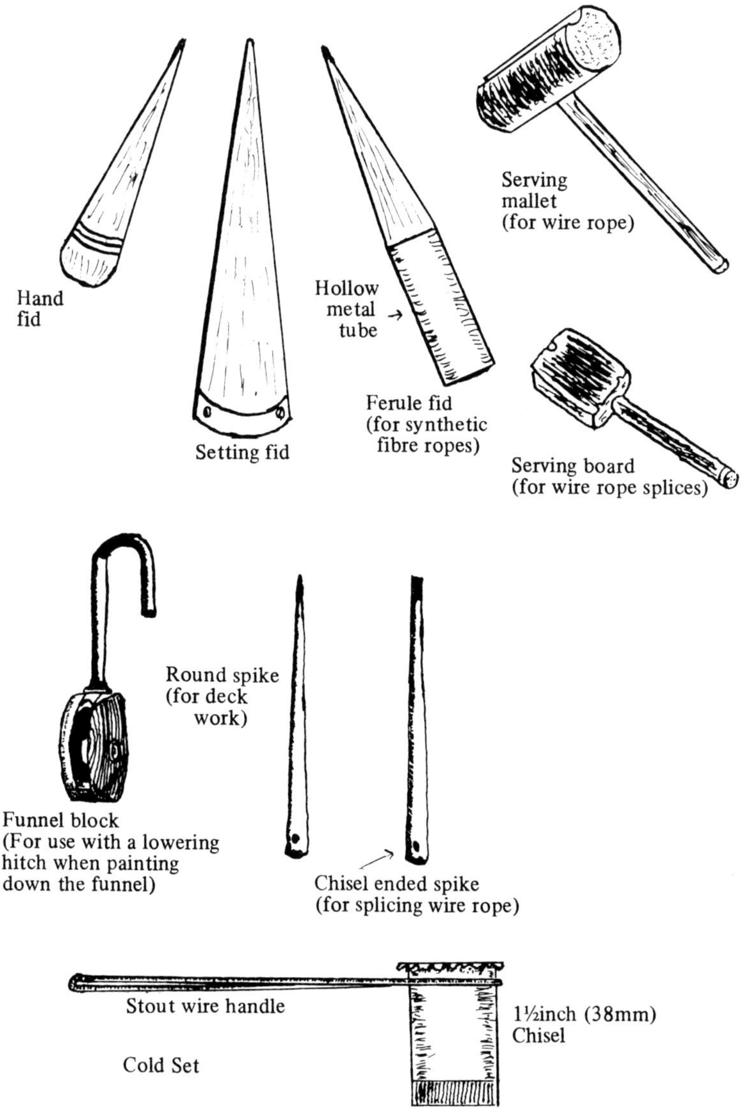
Hand fid
Setting fid
Hollow metal tube
Ferule fid (for synthetic fibre ropes)
Serving mallet (for wire rope)
Serving board (for wire rope splices)
Funnel block (For use with a lowering hitch when painting down the funnel)
Round spike (for deck work)
Chisel ended spike (for splicing wire rope)
Stout wire handle
Cold Set
1½inch (38mm) Chisel

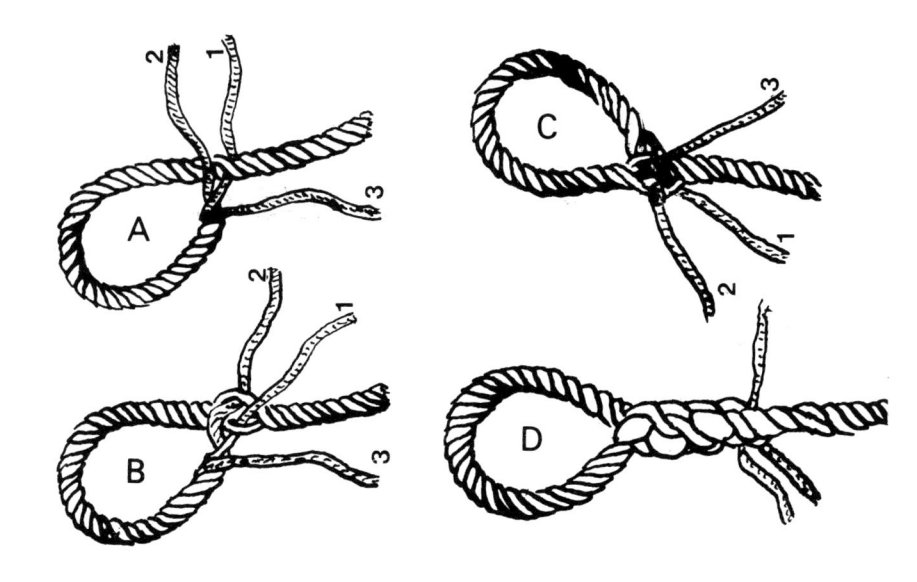

EYE SPLICE

With the eye towards your left, all three strands are tucked from left to right and away from you. Each strand going over one strand and under the next strand, against the lay of the rope.

over one strand and under the next, against the lay. Give each tail a second tuck, pull all tightly home and cut off the spare ends.

Normally the end of a rope is whipped with a palm and needle whipping but back splices are used on the ends of heaving lines and similar tackle.

Eye splice. With the tails of the rope prepared, the centre tail as shown in Fig. A. is that which will be tucked first.

1. With the eye towards your left, insert the fid against the lay of the rope and open up the bight of the strand adjacent to the centre tail. Pass the centre tail through the opened bight against the lay (away from you).
2. Pull the centre tail tight home and open up the bight of the strand to the left of it, with the fid. Insert the left hand tail under the opened strand against the lay (away from you) as in Fig. B. and pull tight.
3. Turn the splice over and open up the remaining bight with the fid. Pass the remaining tail through the bight against the lay (away from you) Haul each tail tight home to ensure that the splice fits snugly, as in Fig. C.
N.B. Each tail should now be issuing from a separate place in the rope. If two tails come out of the same place, the splice is incorrect.
4. Each tail is now tucked over one strand and under the next strand, against the lay (away from you). Continue tucking each tail over and under one strand until the required number of tucks are completed. Five full tucks are required when splicing synthetic fibre ropes and three full tucks for a natural fibre rope.
5. When the tucking is completed, as in Fig. D., unlay each of the tails and divide it into two. Each half is matched with its partner of the next tail and the two halves are seized together over the intermediate strand. The surplus tails are cut off and the ends of synthetic fibres are then heat sealed.

Short splice. When the tails of each rope are prepared, marry the two ropes so that the tails of each rope come out alternately. Seize one set of tails onto the rope they are going to be tucked into. Cut the whipping on the other rope. With the standing part toward your left, tuck each tail over one strand and under the next against the lay. Pull all the tails tightly home, then in the same manner, tucking a tail at a time in rotation, give each tail two more full tucks, over one strand and under the next, against the lay, for a natural fibre rope. Give each tail four more full tucks in the same manner for a synthetic fibre rope. Turn the rope and after removing the whipping, tuck the remaining three tails in the same manner. Pull all the tails tightly home and finish each end of the splice by dogging the tails in the same manner as for an eye splice.

There is no real need to halve the tails and taper a fibre rope splice. The splice will bed down with use.

Apart from putting a good whipping on the tail of each strand, and using some method of preventing the yarns of a synthetic fibre rope from partially unlaying. A practised hand will often dispense with many of the various seizings and whippings suggested in the foregoing instructions. However, they can be of considerable help to a learner, inasmuch as they prevent the end of the rope from becoming unlaid, so avoiding confusion and the waste of good rope. The most important thing is to get each set of tucks correctly and tightly tucked, before proceeding with the next.

TO SPLICE SQUARELINE OR MULTIPLAIT ROPE.

Measure off a length of rope from the end, equal to about five times the circumference of the rope and put a good tight whipping on the rope at this

SHORT SPLICE

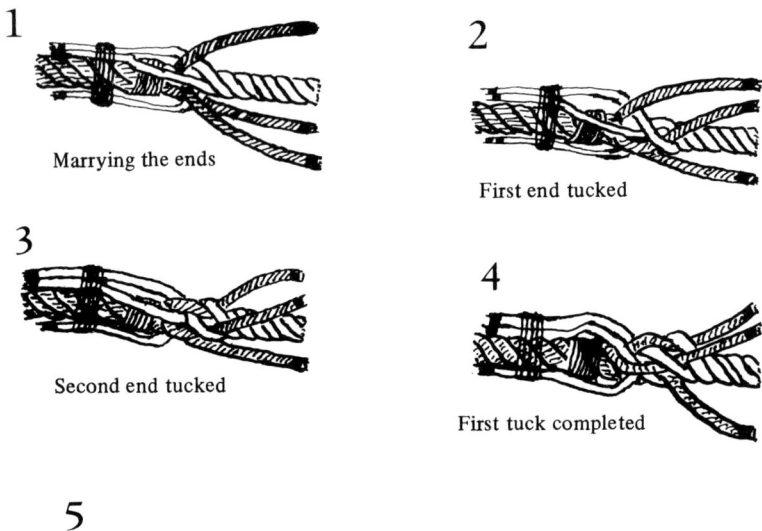

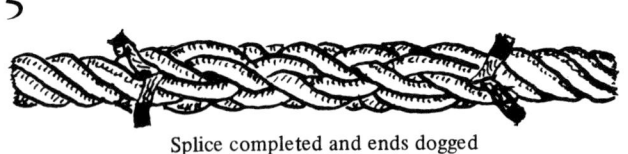

1 Marrying the ends

2 First end tucked

3 Second end tucked

4 First tuck completed

5 Splice completed and ends dogged

CUT SPLICE

HAWKINS MULTIPLAIT

THE 8 STRAND
MULTI-PURPOSE
MOORING ROPE
FROM HAWKINS & TIPSON
ROPEMAKERS LTD

SHORT OR BUTT SPLICING illustrated Step by Step

FIRST STAGE

1. Whip both ropes 2 ft. from the end, then split each into 8 strands, paired as follows: A1/A2—'S' twist; B1/B2—'Z' twist; C1/C2—'S' twist; D1/D2—'Z' twist. Position ropes so that it is possible to bring one pair of 'S' twist strands of each rope from the bottom up to the top as in the diagram. Crossed circle indicates position of Splicer.

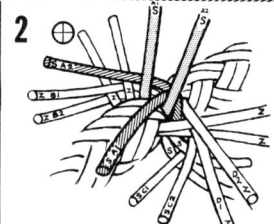

S Z

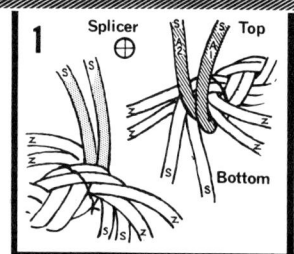

2. In the next movement the ropes are married; 'S' twist Pairs with 'S' twist Pairs, and 'Z' with 'Z'. First interlace A1/A2 (shaded with lines) with A1/A2 (shaded with dots) *exactly as shown— this is most important.*

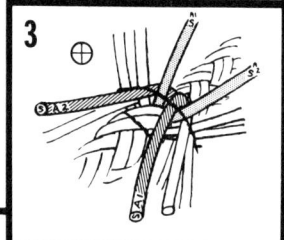

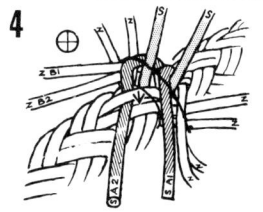

3. Follow round and interlace the remaining strands. When all 16 strands are interlaced, tie as shown.

SECOND STAGE

The movements in this stage are exactly the same as those in figures 8 to 15 of the EYE-SPLICING FOLDER. These movements are made with single strands.

AS FIG. 8 IN EYE SPLICE FOLDER

4. A2 strand (shaded with lines) is tucked UNDER ONE STRAND ONLY and away from Splicer, as shown. Arrow shows direction and strand under which A1 is now tucked, as in 9.

Quarter-turn the rope away from Splicer and tuck B1 strand as shown in 10, then tuck B2 as shown in 11. Quarter-turn rope away from Splicer and tuck C2 as in 12, and C1 as in 13. Quarter-turn rope away from Splicer again and complete tucks as in 14 and 15.

After completing all tucks with one rope, repeat these movements with the other rope, *then cut ties and draw splice tight from each direction.* REPEAT ALL MOVEMENTS IN 8 TO 15 THREE TIMES WITH EACH ROPE, THUS GIVING FOUR TUCKS IN EACH DIRECTION. To finish off, whip ends in pairs—e.g. A1/A2 and B1/B2 etc.

THE COMPLETED SPLICE . . .

showing path of A1/A2 down each rope.

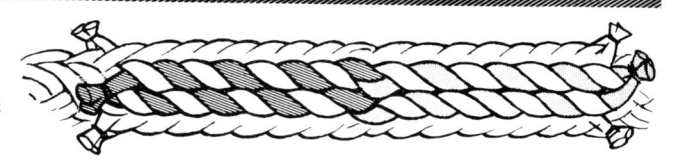

FIRST STAGE
Throughout the first stage, all movements are made with *PAIRS* of strands, i.e. A1/A2, and B1/B2, and so on . . .

1 Whip rope and split end into 8 strands in pairs as follows:

A1/A2 — 'S' twist
B1/B2 — 'Z' twist
C1/C2 — 'S' twist
D1/D2 — 'Z' twist

. . . . and form loop to required size as shown. Crossed circle indicates position of Splicer. The arrow marks the 'S' twist pair of strands (matching A1/A2) and the direction in which the first tuck will be made.

2 Fid is passed through 'S' twist pair, marked by arrow. A1/A2 pair, shaded red, is then tucked through in direction of arrow.

3 Illustration shows first tuck completed. Arrow marks one of the strands that B1/B2 will tuck under.

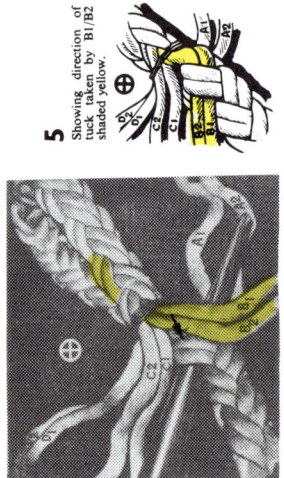

4 Turn rope a quarter turn away from the Splicer. Insert Fid through the 'Z' twist pair and tuck strands B1/B2 (shaded yellow) in direction of arrow.

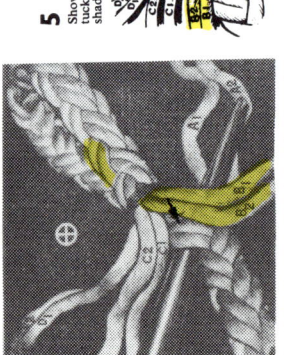

5 Showing direction of tuck taken by B1/B2 shaded yellow.

6 Again quarter turn the rope away from Splicer, and tuck C1/C2, shaded red, with movements as A1/A2 in 2. The strands, through which D1/D2 will tuck, are arrowed.

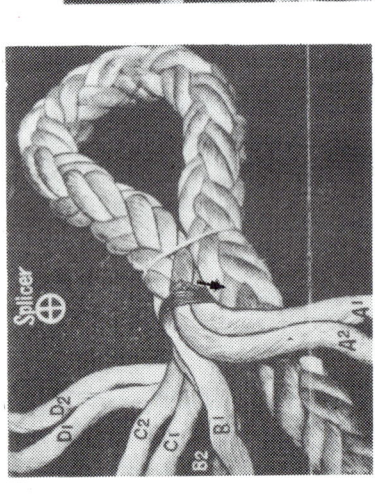

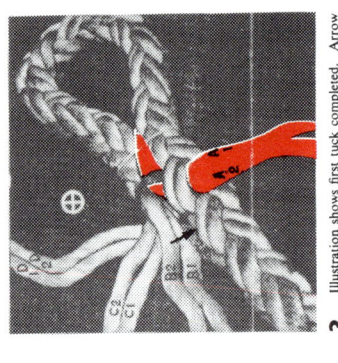

7 Quarter turn the rope away from Splicer once more. D1/D2, shaded yellow are tucked through as shown, with movements as B1/B2 in 5.

168

SECOND STAGE Throughout this stage, all movements are made with single strands *NOT* with pairs

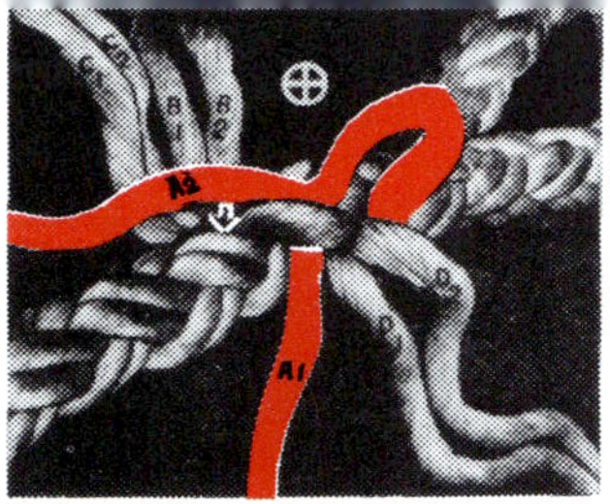

8 Position the job as 2. A2 strand, in red, *nearest Splicer*, is tucked *UNDER ONE STRAND ONLY* and away from Splicer, as shown. Arrow shows strand and direction under which A1 will tuck.

9 Retain same position as 8. Insert A1 under strand arrowed in 8. Note position of A2. Arrow points to strand through which A2 will tuck on the first repeat of movements 8 to 15.

10 Quarter turn the rope from Splicer. B1 strand (shaded yellow) farthest from Splicer, to be tucked back under strand towards Splicer. Arrow points to the strand under which B2 will tuck.

11 Tuck B2 (shaded yellow) under strand arrowed in 10. Note position of B1.

12 Quarter turn rope away from Splicer. Taking strand C2 repeat all movements as with A2 in 8.

13 Taking strand C1, repeat all movements as with A1 in 9.

14 Quarter turn rope, away from Splicer and with strand D1 repeat all movements as with B1 in 10.

15 Taking strand D2 repeat all movements as with B2 in 11.

REPEAT ALL MOVEMENTS 8 to 15 TWICE, THUS GIVING A FOUR TUCK SPLICE.

To finish off, whip ends in pairs A1/A2, B1/B2, C1/C2, D1/D2.

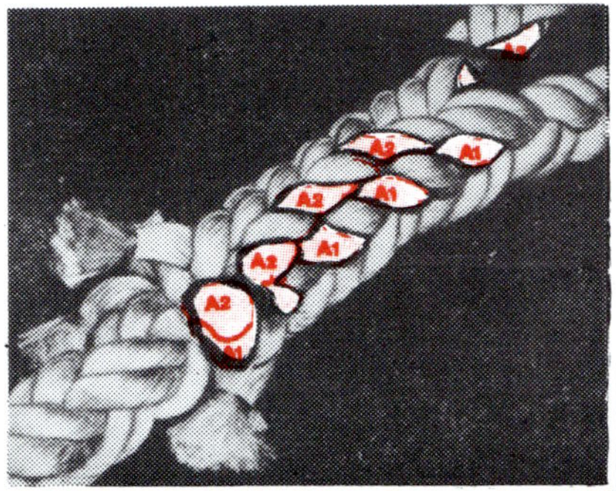

THE COMPLETED SPLICE

... showing path of A1/A2 down the rope.

Note the first movement of paired strands, and then six movements of single strands terminating in whipped ends.

HAWKINS MULTI-PLAIT 8 STRAND ROPE
SPLICING *Illustrated Step by Step*

SQUARE LINE EYE SPLICE

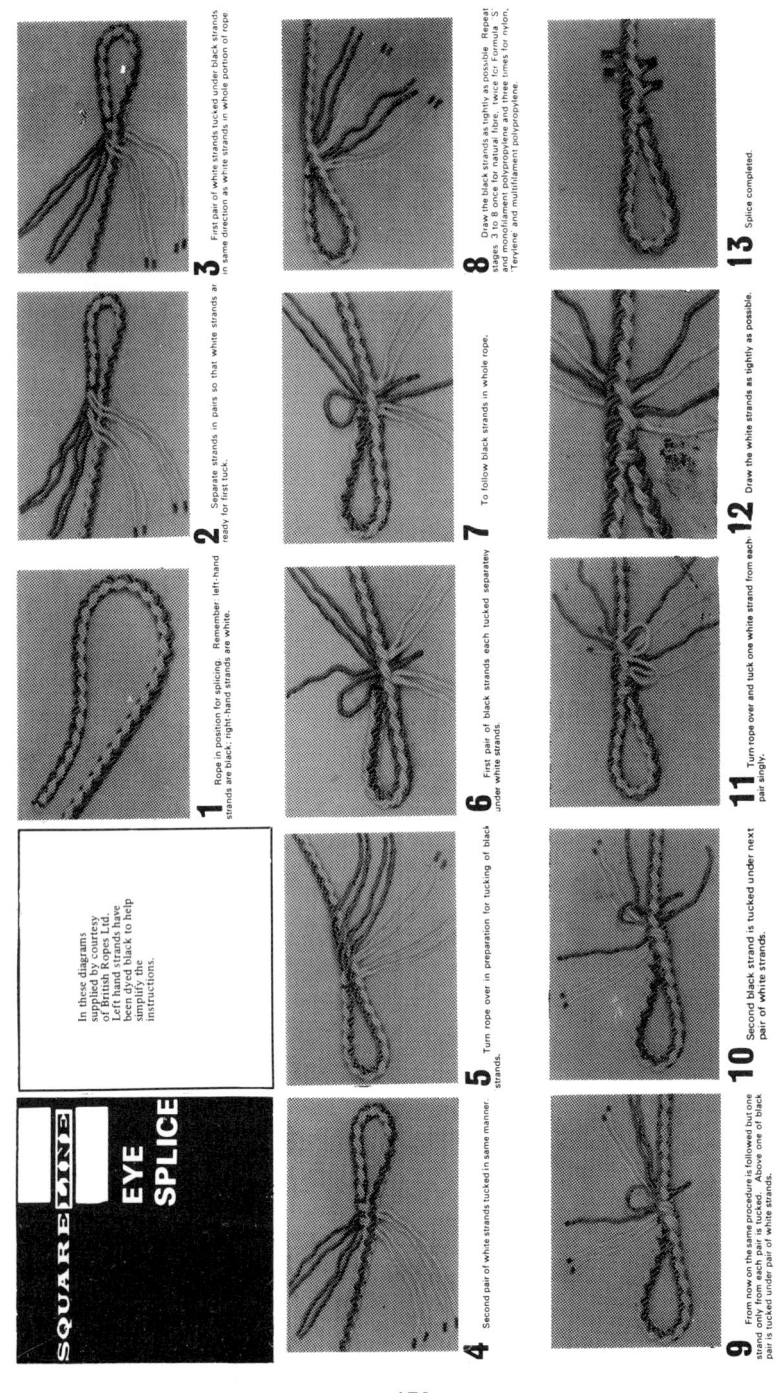

In these diagrams supplied by courtesy of British Ropes Ltd. Left hand strands have been dyed black to help simplify the instructions.

1 Rope in position for splicing. Remember left-hand strands are black, right-hand strands are white.

2 Separate strands in pairs so that white strands are ready for first tuck.

3 First pair of white strands tucked under black strands in same direction as white strands in whole portion of rope.

4 Second pair of white strands tucked in same manner.

5 Turn rope over in preparation for tucking of black strands.

6 First pair of black strands each tucked separately under white strands.

7 To follow black strands in whole rope.

8 Draw the black strands as tightly as possible. Repeat stages 3 to 8 once for natural fibre, twice for Formula 'S' and monofilament polypropylene and three times for nylon, Terylene and multifilament polypropylene.

9 From now on the same procedure is followed but one strand only from each pair is tucked. Above one of black pair is tucked under pair of white strands.

10 Second black strand is tucked under next pair of white strands.

11 Turn rope over and tuck one white strand from each pair singly.

12 Draw the white strands as tightly as possible.

13 Splice completed.

170

SQUARELINE
SHORT SPLICE

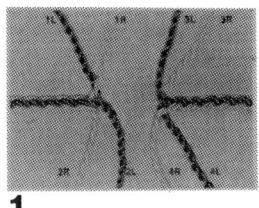

1 Unlay rope and separate pairs of strands, tying the ends of pairs together. In the photograph the left-hand strands have been darkened, and the right-hand strands are light. The pairs are designated from 1 to 4L meaning left-hand twist, and 1 to 4R meaning right-hand twist.

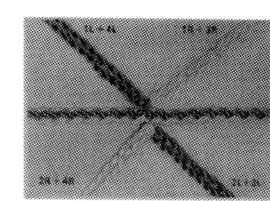

2 Now the strands are to be married and if this is done correctly no difficulty should be experienced.

Tie Together: 2L and 3L   1R and 3R
              2R and 4R   1L and 4L

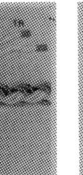

3 Nos. 1R to follow 3R in same direction as white strands in whole portion of rope.

4 Repeat with 1R – 3 tucks (natural fibre). For man-made fibre, repeat with 1R – 4 tucks.

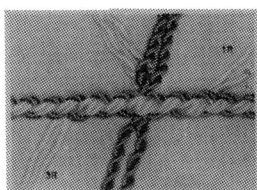

5 Nos. 3R to follow 1R in opposite direction, to complete 3 tucks.

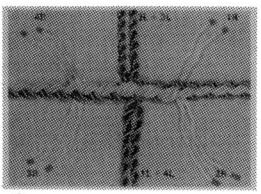

6 First stage of $\frac{1R \text{ and } 3R}{2R \text{ and } 4R}$ completed

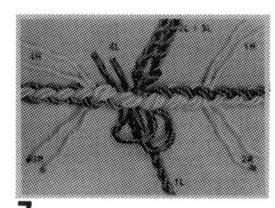

7 Repeat stages 3 to 6 with $\frac{1L - 4L}{2L - 3L}$ Three tucks in opposite directions.

8 Splice complete.

USING A PALM AND NEEDLE.

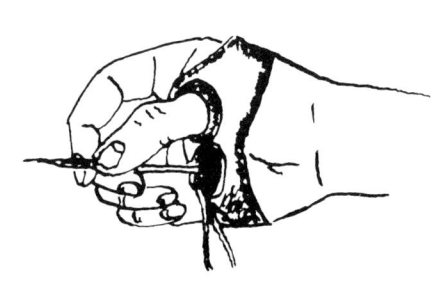

Sail and roping needles are triangular.

Roping needle sizes 6 - 12
Sail needle sizes 12½ - 16
Low numbers - Big needles.
High numbers - small needles.

point. Unlay the strands as far as the whipping and whip the tails tightly. With synthetic fibres the tails may be heat sealed but as the fibres are much finer than natural fibres, the tails of synthetic ropes should in addition be whipped about every six inches (15cm), or laced, in order to ensure that they do not become partially unlaid.

Form an eye of the size required at the end of the rope, leaving the tails clear of the eye and seize the end of the rope onto the standing part.

Two pairs of the strands of the rope are laid up right-handed and the other two pairs are laid up left-handed so that the lay forms a plait.

Mark each of the four left-hand laid tails by some distinctive means, such as passing a strand of coloured wool or thread through each tail, near the end. This ensures that as the splice progresses, the tails are not confused. If the tails are allowed to become confused, the splice becomes impossible.

In the diagrams supplied by courtsey of British Ropes Ltd. left-hand strands have been dyed black to help simplify the instructions.

TO SPLICE FLEXIBLE STEEL WIRE ROPE. (six stranded)

Splicing wire rope is always much easier if the tools of the trade are to hand. It is however most unlikely that a modern ship will be equipped with a rigger's vice or even a rigging scew. An engineer's vice situated so that it is possible to hang the wire above the vice is the next best thing. Failing that, the wire will have to be stretched between two stanchions and spliced horizontally. Although splicing a wire in this manner is quite a simple matter, the lack of any firmness makes it rather difficult for a beginner.

As always, tools first, seaming twine, marline, a sharp knife, two chisel ended spikes of suitable size and if a thimble is to be spliced in the eye, a suitable thimble. A 2 lb. (.9kg) hammer and if the wire is to be cut before splicing a heavy hammer and cold set.

When the wire is to be cut before splicing. Decide on the place where the wire is to be cut and put two strong whippings on the wire, about an inch (2.5cm) apart, so that the wire may be cut between them. Use a heavy hammer and cold set to cut the wire.

Always have two spikes handy, the second spike is used to pick up any wires that the first spike has missed or to drop any extra wires the first spike may have picked up.

The following together with the diagrams is an extract from British Ropes Ltd. Technical information bulletin No. 4. and gives directions for the making of a wire snorter with a thimble eye each end. When splicing a soft eye simply disregard the instructions for inserting the thimble. The splice is suitable for any wire rope aboard the ship.

Preparation

All wire ropes need some preparation before splicing commences and, apart from a variation in the length of the tucking strands or tails for some work, this preparation is normally as follows.

It will be assumed that a rope is required to be spliced at each end to a given length, having a thimble at each end. Measure up the rope for the required length and mark where it is to form the centre of the crown of the eyes, making due allowance for the loss in the straight length caused by the curving of the rope around the thimbles.

WIRE ROPE FITTINGS

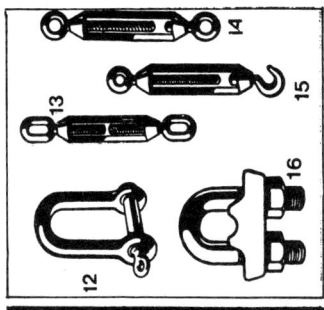

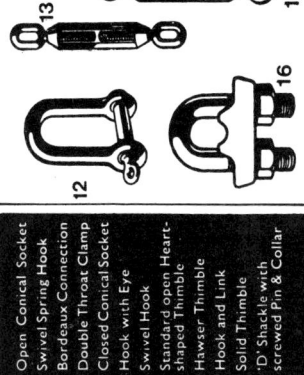

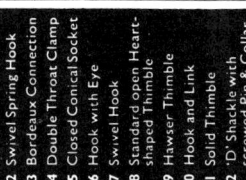

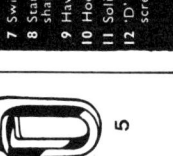

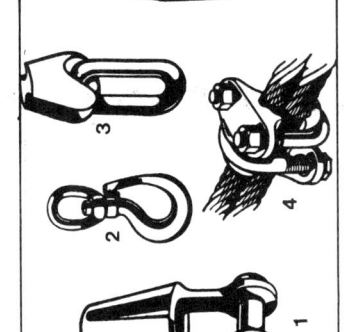

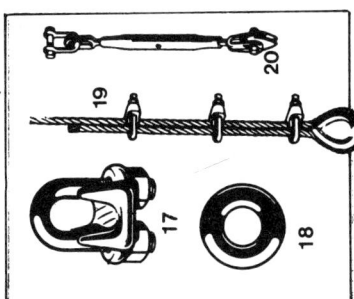

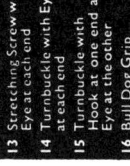

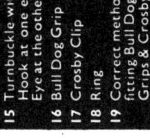

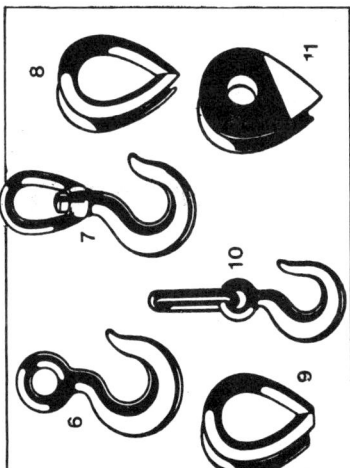

1. Open Conical Socket
2. Swivel Spring Hook
3. Bordeaux Connection
4. Double Throat Clamp
5. Closed Conical Socket
6. Hook with Eye
7. Swivel Hook
8. Standard open Heart-shaped Thimble
9. Hawser Thimble
10. Hook and Link
11. Solid Thimble
12. 'D' Shackle with screwed Pin & Collar
13. Stretching Screw with Eye at each end
14. Turnbuckle with Eye at each end
15. Turnbuckle with Hook at one end and Eye at the other
16. Bull Dog Grip
17. Crosby Clip
18. Ring
19. Correct method of fitting Bull Dog Grips & Crosby Clips
20. Rigging Screw

Other designs of fittings including those of the hand-forged type, can be supplied and attached to ropes, if desired.

Place the thimble with its crown on this mark and bend the short end of the rope around and in the groove of the thimble, leaving the short end of rope protruding there from, and which will provide the length for the tucking tails. (This length should be 4 inches for each $\frac{1}{8}$th inch of the rope's diameter except in the case of a splice with a Bordeaux connection when the tails should be 7 inches for each $\frac{1}{8}$th inch of the rope's diameter). (N.B. Equal to one foot of wire for every inch of circumference for an ordinary splice. Ed.)

Force the two parts of rope close together at the thimble points and where the short end leaves the thimble points mark the rope (B).

If the short end is longer than is necessary for the tails, measure off the length required and mark (C). Remove the thimble.

At mark (B) firmly bind the rope with fine fibre, working away from the short end for a distance equal to two diameters of the rope. If necessary put cutting bindings on each side of mark (C) and cut rope.

Re-assemble the thimble at its appropriate position, bend the rope around and force the two parts of rope together at the points of the thimble, i.e., the crutch of the eye. Firmly sieze these two parts together with strong wire—they should be in direct contact. This seizing should only consist of two or three turns of wire.

Prepare the other end in a similar manner and check the length of the assembly.

The rope and thimble should be tightly bound together at the crown of the thimble and again on each flank, serving towards the point of the thimble, and as close as possible thereto.

This ensures that the rope is in solid contact in the groove of the thimble and that the two parts of rope are together at the crutch. The end is then ready for splicing.

(In those cases where a rope is to be spliced at one end only and the length does not matter to an inch or two, the preparation can start from the rope's end. Measure back for the lengths of the tails, reeve on binding "B", assemble with the thimble and sieze up).

Place the thimble in a vice with the rope leading vertical and the short end, i.e., the "tails" end, on the left hand side.

Remove end binding and unlay the short end of rope to provide the tails for splicing, and remove the crutch seizing. (See Fig 1).

The fibre main core is to be tucked into the main part together with tail No. 1, for the first tuck (see diagrams), it will then be cut off where it emerges from the main part.

With ropes made with a wire core, the wire core **must never be cut from the rope.** It must be split up, and the wires or strands distributed among the tucking tails, and tucked with them for at least three tucks.

If the rope is not Preformed it is advisable to bind the ends of each tail separately.

In all splices the spike must be entered as near as possible to the thimble or end fitting, and the tucking tail must enter into that portion of loop so formed which is nearest the thimble or end fitting, i.e., under the spike. All tucks must be pulled down hard.

To "Break out" wires when reducing the number of wires per strand, take each wire separately, snatch back to the point where it emerges from the rope and then twist wire (handle fashion) reversing direction if necessary and the wire should part in the gusset.

SIX STRANDED ROPE: FIVE TUCK SPLICE

THIMBLE in VICE. ROPE Vertical. MAIN part of ROPE on RIGHT hand. TAIL strands on LEFT hand.
THIMBLE seized at CROWN and both flanks.
STRANDS FOR THE TAILS separated and whipped at ends.
LENGTH of TAILS for a FIVE TUCK SPLICE 4" for each $\frac{1}{8}$" diameter of rope.

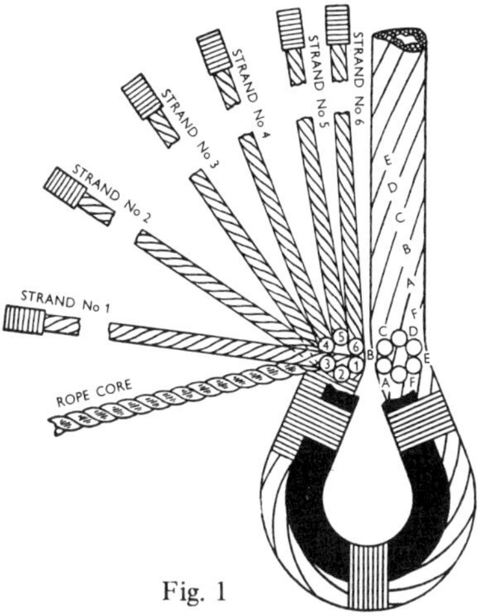

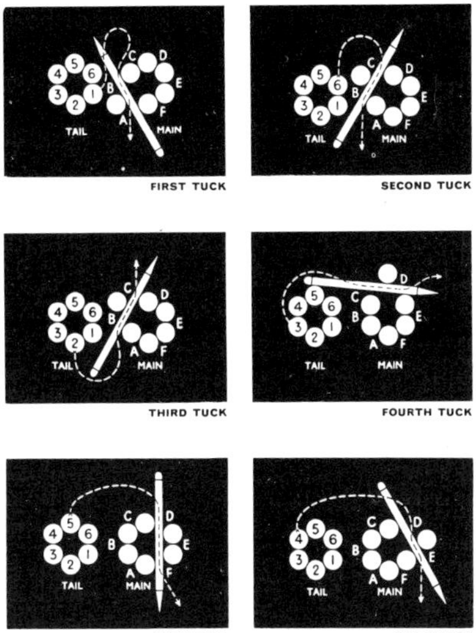

FIRST TUCK

SECOND TUCK

THIRD TUCK

FOURTH TUCK

FIFTH TUCK

SIXTH TUCK

Fig. 1

Diagram showing emergence of tails after the First Series is completed.

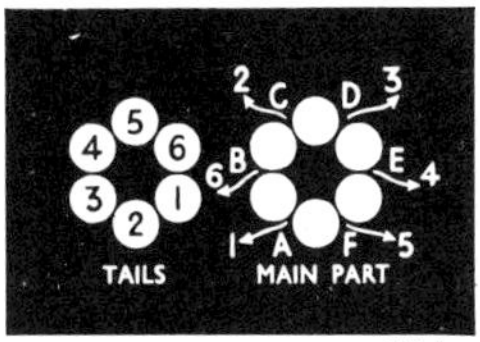

FIG. 2

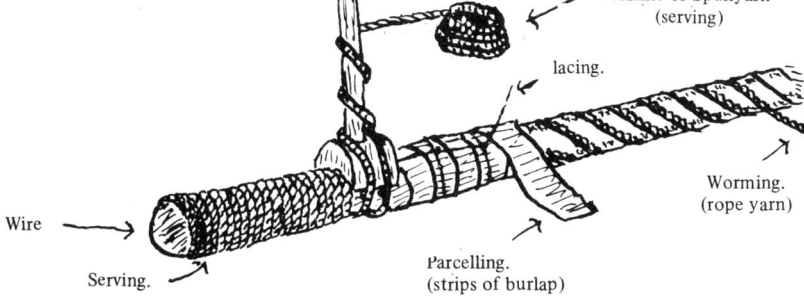

WORMING - PARCELLING - and SERVING.

"Worm and parcel with the lay -
Turn and serve the other way."

The serving mallet (or board) is used to get the marline tight and the ball must be passed round and round the wire by an assistant as the operator serves the wire.

175

The following method of splicing would be in accordance with the Shipbuilding and Ship repairing Regulations 1960 Part IV, Regulation 39 and Docks Regulations 20(d).

First Series of Tucks

A **Fibre** Main Core should be tucked with tail No. 1 and then cut off. A **Wire** Main Core must be split up, distributed among the tails, and tucked with them for at least three series.

First Series

Tail No.	In at	Out at
1	B	A
6	C	B
2	B	C
3	C	D
5	D	F
4	D	E

Note:- When tucking tail No. 3, pass it inside tails Nos. 4 and 5 to tuck it under strand D and NOT outside as outlined in the diagram of No. 4 tuck.

Second Series

Tail No.	In at	Out at
1	B	C
6	C	D
2	D	E
3	E	F
4	F	A
5	A	B

Third Series

In at	Out at
D	E
E	F
F	A
A	B
B	C
C	D

When tucking tails, always ensure that the tail about to be tucked passes inside and NOT outside the next tail for tucking.

After the Third series, the wires of a wire main core may be "broken off", and the number of wires in each of the main tails reduced to half the original number, preferably by "breaking out". The remaining wires to be twisted up to a rough strand formation, and at the same time enclosing the cut ends in the centre thereof.

Fourth Series

Tail No.	In at	Out at
1	F	A
6	A	B
2	B	C
3	C	D
4	D	E
5	E	F

Fifth Series

In at	Out at
B	C
C	D
D	E
E	F
F	A
A	B

Remove splice and hammer down the taper, starting from the eye and working down the taper. This is to tighten up the tucks and to round up the taper. Remove protruding wire ends, preferably by breaking out, and again round up over the broken off ends. The taper (or at least that portion containing the wire ends of the tails) should be served with wire strand or spunyarn to give protection to the user when handling.

The need for thoroughly pulling down each strand as tightly as possible as splicing proceeds, cannot be over-emphasised. The tails should be pulled down in line with the centre line of the thimble. To get the tuck tight and short, it should be beaten by means of a mallet or hammer. One object is to get the tuck as nearly as possible at right angles to the axis of the rope. Working the tucks with mallet or hammer forces any slackness out of the tucking tails through the loop, and the beating should start on the position of the tail before its entry into the rope, and continue to the tuck itself. The strands of the main rope where they have been lifted are beaten down to hold the tuck in place.

In addition to the foregoing instructions. It always helps to keep a neat splice, if as each series of tucks is completed, each tail is hammered well home and a tight seizing or gag is placed round the tucks. This prevents the tails loosening in the wire while the next series of tucks is being made.

Every care must be taken not to allow the tails to kink as they are being tucked. If a tail has a tendency to kink, do not try to pull it home hoping the kink will put out, it will not. Take the tail out twist it to take a half turn out and retuck, being very careful not to allow the kink to reform in the bight of the tail.

Splices made with the lay of the rope, are another form of splice, but experience in use has shown that they are not as efficient as the form of splice that contains a locking tuck. Therefore splices made in wire ropes with the lay, should never be used on any wire rope when its end or ends are free to twist or on wire rope slings or sling legs. They are however a perfectly safe splice to use on mooring wires and wire cargo lashings.

SERVING.

When the splice is completed it should be served:—
(a) To help protect it from corrosion.
(b) To protect the operators hands from being torn by the bare ends.
(c) To give a neat and tidy appearance.
To serve a wire properly it must first be well greased, then wormed by laying strands of rope yarn in the lay of the rope to fill the vacant spaces. The whole is then wrapped in strips of burlap, this is parcelling. Both the worming and parcelling are done with the lay. The parcelling is then kept in position with a lacing of seaming twine or marline, to prevent it loosening up while the operation of serving is carried out.

Worm and parcel with the lay.
Turn and serve the other way.

With the aid of a serving mallet (or serving board when a splice is being served) the wire is tightly bound against the lay with either marline or spunyarn. When serving a splice, always serve from the standing part towards the eye. Start and finish serving in the same way as a common whipping is started and finished, making fast the end with a clove hitch on the serving turns to prevent it coming adrift.

CHAPTER 10.

LIFTING TACKLE—BLOCKS, PURCHASES, MASTS, DERRICKS, CRANES, WINCHES.

Blocks. The word block when referred to in a dictionary, has a multitude of meanings. In marine circles, a block consists of a pulley wheel or wheels confined in a binding (frame). These blocks are portable and normally used in pairs to assist in the transporting of a heavy object from one position to another, or singly to change the direction in which the rope moving the object is leading.

The sheaves (pulley wheels) may be made of wood (lignum vitae), galvanised iron, or cast iron, depending upon the type of block to which the sheave is fitted. The centre of the sheave is fitted to a pin, around which it is able to revolve and to which some form of lubrication must be applied. In some wood sheaves a hole to take the pin is punched out, normally however, a bush will be fitted into the centre of the sheave to take the pin, this bush can be plain but more often contains roller bearings.

The binding is made of steel and may have either a wood or steel shell inside, or outside, or no shell at all. In some wood blocks (clump blocks) the binding is dispensed with, the shell being scored to take a wire or fibre rope strop which is used to anchor the block.

All blocks are stamped with the S.W.L. (Safe working load) and or the rope size for which they are suitable.

Parts of a block.
Arse.   That end of a block which does not contain the swallow.
Becket.   One fork of the binding may be extended at the arse, for the purpose of attaching the standing part of the fall to the block.
Binding.   Steel forked piece, one end of which is fitted to secure the block to a strong point, while the forks accommodate the pin. (one fork may be extended to make the becket).
Bush.   A centre piece in the sheave which revolves around the pin.
Cheek.   A side of the shell of the block.
Crown.   That end of the block which houses the swallow and the part of the binding used to secure the block.
Distance pieces. Pieces used at the arse and crown to keep the cheeks of the shell apart and ensure free running of the sheave or sheaves.
Pin.   Axle upon which the sheave revolves.
Saucer.   Heavy steel saucer placed immediately below the crown of the floating block on some heavy lift purchases for the purpose of pulling the lifted weight in a desired direction, causing the derrick to follow.
Score.   Groove cut in the shell of clump blocks for the purpose of holding the strop in place.
Sheave.   Pully wheel around which the fall is led.
Shell.   A cover over the sides, top and bottom of the sheaves.

Types of Head Fittings

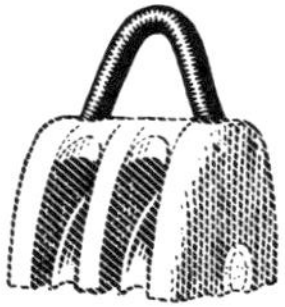

BOW

SWIVEL HOOK

SWIVEL RING

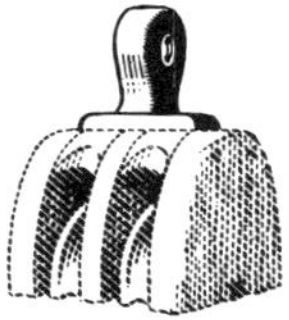

STUD or SOLID EYE

DOUBLE LUGS

SWIVEL EYE & LOOSE LINK

SPECIAL TYPES

Made with any fittings required

WOOD SNATCH BLOCK

WOOD NON-TOPPLING BLOCK

WOOD BLOCKS FOR ROPE STROP

SINGLE

DOUBLE

SHEAVES FOR WOOD BLOCKS

Lignum Vitae Patent Sheave

Galvanised Iron Patent Sheave

Lignum Vitae Plain Sheave

Galvanised Iron Plain Sheave

WOOD PULLEY BLOCKS—Internal Bound
WITH SWIVEL OVAL EYE

SINGLE With Becket

DOUBLE With Becket

TREBLE With Becket

SINGLE Without Becket

DOUBLE Without Becket

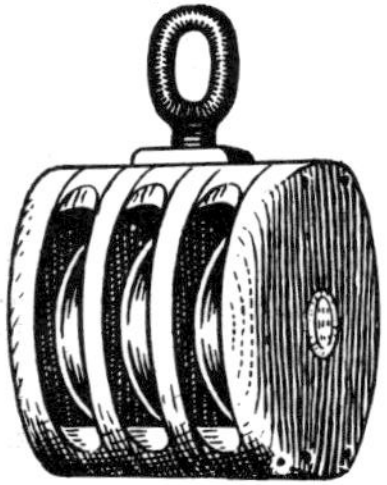

TREBLE Without Becket

WIRE ROPE BLOCKS
FOR SHIPPING & ENGINEERING

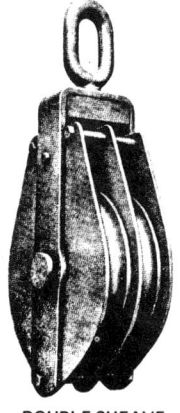

SNATCH BLOCK

ROLLER BEARING BLOCK

'Z' TYPE CARGO BLOCK

DOUBLE SHEAVE
SWIVEL OVAL EYE

GIN BLOCK

TREBLE SHEAVE
SWIVEL OVAL EYE

LOWER
PURCHASE BLOCK

70 TON LOWER PURCHASE BLOCK

Strop.	A grommet around a block that has no binding. Used to secure the block to a strong point.
Swallow.	The point in the shell where the fall enters or leaves the block as it passes over the sheave.
Swivel.	A pivot included in the binding so that the block may turn when it is secured to a strong point.
S.W.L.	Safe working load.
Tail.	A fibre rope tail affixed to the binding or strop of a small block for the purpose of securing the block to a strong point.

Types of blocks.

Cargo block.	External bound metal block used on a derrick with a wire rope fall as head block, heel block, topping lift block or purchase block.
Clump block.	Wood block with no binding but with the shell scored to take a strop. Originally the shell was carved from one piece.
External bound block.	A block having the forks of the binding outside the shell.
Funnel block.	A block constructed with a long hook at the crown, so that it may be hooked over the funnel top, for use when painting the funnel.
Gin block.	Metal block with a skeleton binding and no shell. Normally used for working cargo with fibre rope whips.
Internal bound block.	Wood block with the forks of the binding enclosed in the shell.
Metal block.	Any block made entirely of metal.
Non-toppling block.	A block so constructed that when used as a floating block on a lifeboat fall or cargo purchase, the crown will remain below the arse and the block will not topple when no weight is suspended from the purchase.
Snatch block.	A block so made that one cheek or a part of one cheek is hinged and allows a fall to be placed in the swallow, without having to reeve the fall.
Wood block.	Any block having a wood shell.

Care of blocks. It is essential that all blocks are taken adrift, overhauled and lubricated at regular and frequent intervals.

To overhaul a wood block, remove the plate stating the S.W.L. and or rope size, which is nailed over the head of the pin. Turn the block over and using a metal drift, punch the pin out of the shell. Remove the sheaves and binding. Throughly clean all the parts using paraffin to remove old grease from the sheaves and shell and wire brush the binding. Lubricate the bush, sheave and pin with tallow and black lead or solidified vaseline. Coat the forks of the binding and the inside of the cheeks with a light grease. Wipe the outside of the shell with raw linseed oil. Swivels should be well lubricated with a medium lubricating oil. A swivel that has seized up through lack of attention can usually be freed with the aid of paraffin or diesel oil. In very stubborn cases, wrap a piece of cloth round the swivel, soak with paraffin and set light. Provided the swivel is not strained, the heat will usually do the trick. Be careful not to do this where you may set light to something. When you have put the block together again,

'Dunstos' STANDARD BLOCKS

SECTION THROUGH SHEAVE AND BOSS

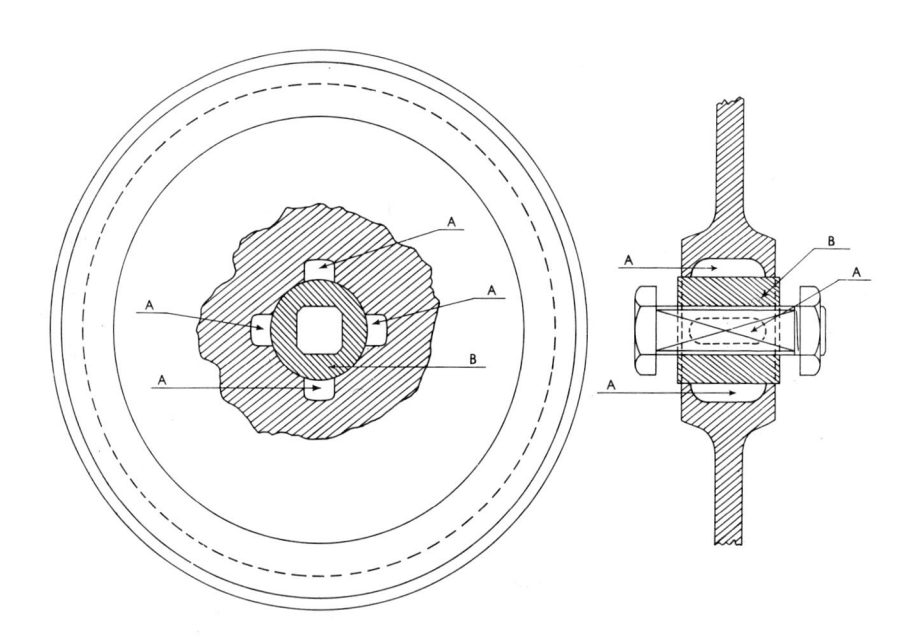

SOME POINTS WORTH NOTING

1st Strength of Centre Pin remains constant throughout life of block as no wear takes place on pin.
2nd Large bearing surface of bush on which the sheave revolves.
On this bolt is mounted cast iron bush 'B' and this comes in direct contact with sheave.
The centre bolt is square with the sharp corners off.
Cavities A.A.A.A. are filled with grease, suitable for extreme hot or cold climates.

DUCK BILL
GOOSENECK FITTING

BOLT "A"
GREASE DOOR
BLOCK CHEEK "B"
NUT "C"
GREASE DOOR

CAST IRON AXIS OR BUSH
BOTTOM OF BUSH OPEN TO ALLOW GREASE TO BE IN DIRECT CONTACT WITH SHEAVE
HOLE FOR SQUARE CENTRE PIN
TROUGH IN BUSH

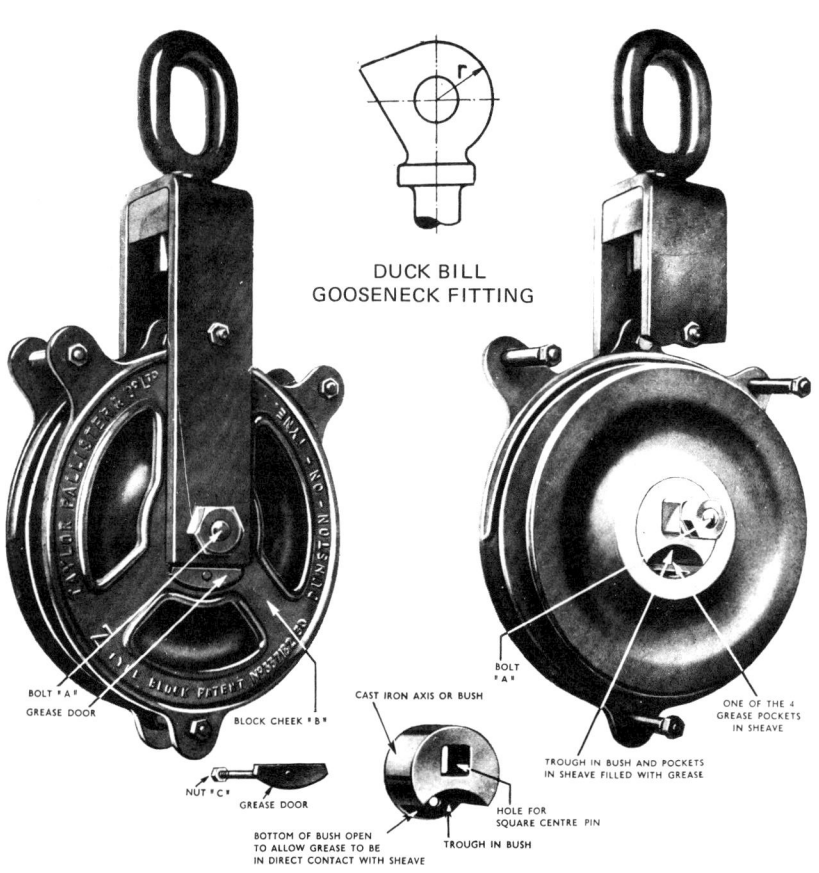

BOLT "A"
TROUGH IN BUSH AND POCKETS IN SHEAVE FILLED WITH GREASE
ONE OF THE 4 GREASE POCKETS IN SHEAVE

'DUNSTOS' TYPE 'Z' CARGO BLOCK

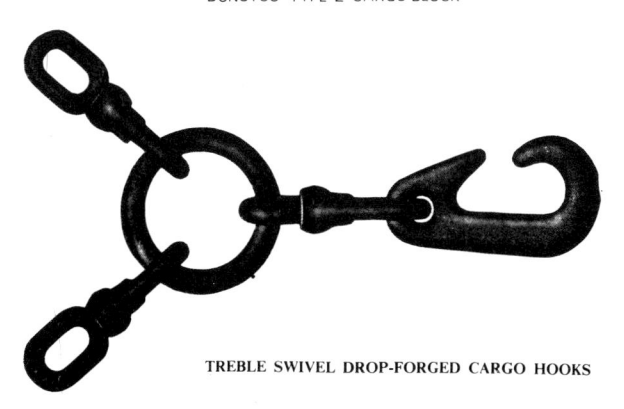

TREBLE SWIVEL DROP-FORGED CARGO HOOKS

don't forget to tack the plate over the head of the pin.

To overhaul a metal block, remove the split pin from the pin and if there is a nut holding the pin in place, undo it. Undo the nuts and bolts holding the cheeks and distance pieces in place at the crown and arse of the block. If the pin has to be punched out, be careful not to damage any thread. Half screw the nut onto the pin and tap the nut lightly with a hammer to punch the pin out. Self-oiling sheaves should be cleaned and the old oil drained out by removing the grub screw. Refill with a medium lubricating oil and replace the grub screw, smear the sheave with blacklead and tallow or a light grease. A sheave which has depressions in the bushes should have the depressions thoroughly cleaned out with paraffin after which they are refilled with solidified vaseline or other suitable solidified grease. Sheaves having a grease nipple should have a suitable grease pumped in until fresh grease is seen to be coming out. Wire brush cheeks and binding, then lightly coat shell, binding, bolts and pin with a light grease, lubricate the swivel if there is one with medium lubricating oil and reassemble. Be careful not to damage the thread on the pin when replacing it. Fit a new split pin and open the end.

With all blocks, always ensure that the sheaves are turning easily after the block has been reassembled and that the swivel is in good working order, before returning the block to it's working position. Any blocks that show signs of scoring on the cheeks or have a bent or damaged pin or a scored or damaged bush in the sheave or on which the swivel appears to be strained, are immediately suspect and should be referred to a competent officer before being reassembled, so that if necessary it can be condemned. On clump blocks examine the strop closely and renew if it shows signs of wear.

Purchases. A purchase consists of a rope rove through two or more blocks, it's purpose being to reduce the weight of the total load at the point where power is being supplied, for the purpose of moving the load.

The mechanical advantage or power gained, will depend on the number of parts of rope at the floating or moving block. With a single whip using only one fixed stationary block, no advantage can be gained and a pull of one ton will just balance a weight of one ton. With a Dutchman's purchase the law is reversed and the mechanical advantage is decreased.

Using a double whip with the standing part of the fall made fast to the fixed block, each of the two parts of the fall coming from the floating block, will bear half the load and a pull of half a ton will balance a weight of one ton. However, using the double whip the other way round, there will be three parts of rope at the floating block (the standing part and two parts coming from the sheave) and a pull of one third of a ton will balance a ton (each part of rope bears a weight of one third of a ton). This method of progression increases as each additional sheave increases the number of parts of rope at the floating block.

In addition when using a purchase, an allowance of ten per cent of the weight has to be added for each sheave in respect of friction, so that the maximum advantage is gained with a four fold tackle. Beyond this point friction takes over and increases the weight of the load. Nevertheless friction can be reduced by increasing the sheave diameter and reducing the rope diameter, by using a smoother rope and efficient lubrication. So that when moving very heavy loads using large sheaves and wire ropes five and six fold purchases can be used with advantage, particularly when both ends of the fall are used as hauling parts.

Wood blocks are normally manufactured for use with fibre ropes only and are measured by the length of the cheek. The length of the cheek should be at least three times the circumference of the fall and there should be ample room in the swallow for the fall to pass through.

Metal blocks can be manufactured to take either fibre or wire falls (but not both). They are measured by the minimum diameter of the sheave and the sheave diameter should be at least sixteen times the diameter or five times the circumference of a wire rope. Blocks built to take fibre ropes will have the rope size stamped on the shell.

To "Fleet a tackle" means to stretch the tackle to its full length.

"Two blocks" means that the two blocks have been hauled together.

A "Thoroughfoot" in the tackle occurs when a block topples over and through the fall.

A "Tail block" is a small block with a fibre rope tail attached.

A "Floating block" is the moving block in a tackle.

Types of purchases or tackles.

Whip.	A fall rove through a single fixed block. M.A. nil.
Runner.	Standing part made fast to a fixed point, single block runs on the bight of the fall. M.A. 2.
Double Whip	Two single blocks. M.A. 2 or 3.
Spanish burton	Two single blocks. M.A. 3. (seldom used).
Gun tackle.	For moving a load horizontally, originally used on Naval vessels for hauling the guns out after the recoil. Two single hook blocks, M.A. 2 or 3.
Handy billy or watch tackle.	Small fibre rope tackle in which the rope size does not exceed 2" (16 mm). One double and one single block. Standing part made fast to the single block, double block has a rope tail attached. M.A. 3.
Jigger.	Same as Handy billy, fall size between 2" (16mm) and 2½" (20mm).
Luff.	Same as handy billy, fall size 3" (24mm) or over for fibre rope. May have wire fall any size. Double block has no tail. M.A. 3 or 4.
Double luff or Two fold.	Two double blocks. M.A. 4 or 5.
Three and two. (Gyn).	One double and one treble block. M.A. 5 or 6.
Three fold.	Two treble blocks, M.A. 6 or 7. Three fold purchases rove with fibre rope falls should have both the standing part and the hauling part rove through the middle sheaves, to avoid tipping the block.
Luff on luff.	Luff tackle made fast to the hauling part of a luff tackle. M.A. rove to disadvantage ÷ 9, rove to advantage and disadvantage ÷ 12, rove to advantage ÷ 16.
Four fold.	Two quadruple blocks. M.A. 8 or 9.
Dutchman's purchase.	Has the weight on the hauling part which is lifted by fleeting the tackle with a tail on the floating block.

Note—M.A.—Mechanical advantage. (no account taken of friction).

Chain hoists. A much more powerful alternative to a purchase is the chain hoist. Whereby the pulling of an endless chain around a gipsy will drive a gearing, which in turn will cause a second gipsy to lift or lower a weight suspended from a second chain. Because the gearing is very low considerable weight can be lifted

Whip   Runner   Double Whip   Spanish Burton

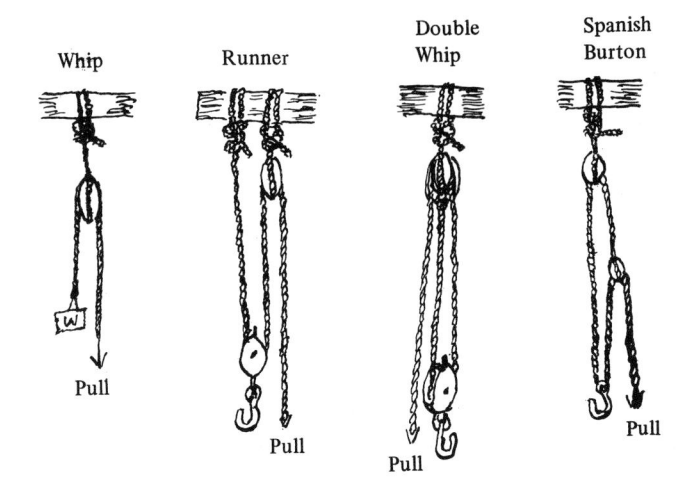

Pull   Pull   Pull   Pull

GUN TACKLE

Pull

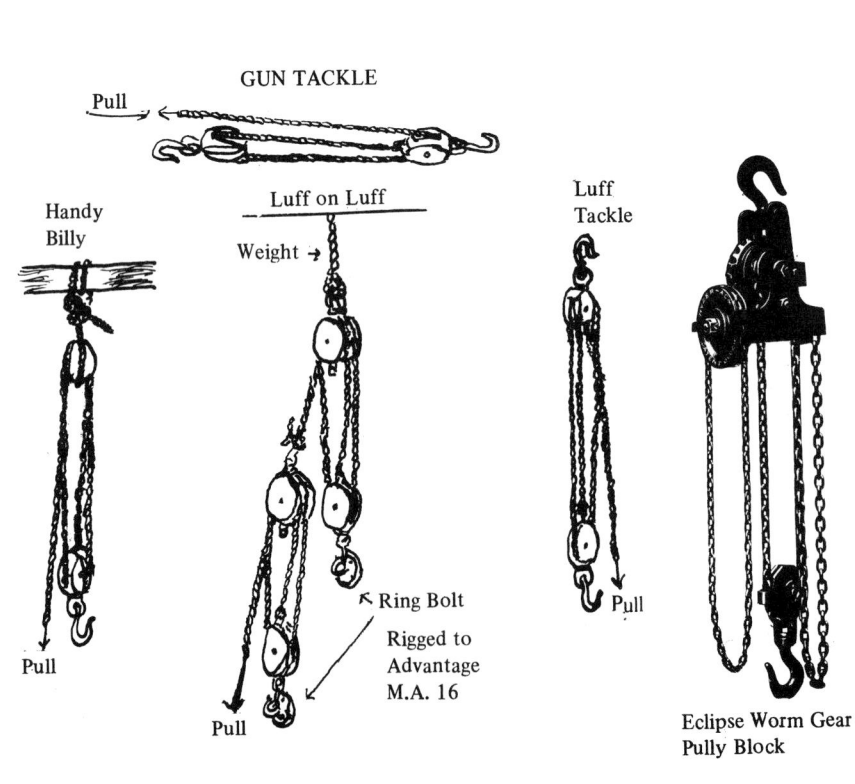

Handy Billy   Luff on Luff   Luff Tackle

Weight →

Pull

← Ring Bolt

Rigged to Advantage M.A. 16

Pull

Pull

Eclipse Worm Gear Pully Block

REEVING A THREE FOLD PURCHASE THROUGH CENTRE SHEAVES

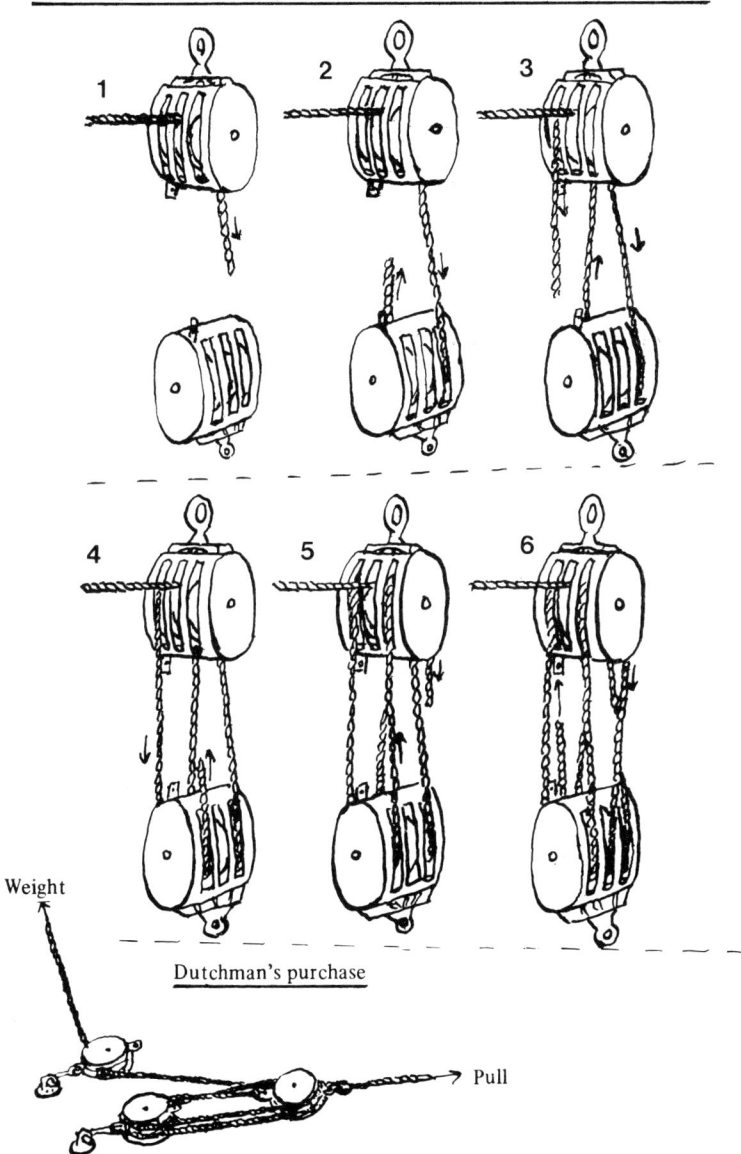

Dutchman's purchase

Weight

Pull

by the sustained application of very little power. A chain hoist is normally available in every ship's engine room, being placed there for the purpose of lifting the cylinder heads and pistons of the main engine.

MASTS.

Masts were originally stepped in ships for the purpose of carrying sails. As sails became redundant, the masts continued to serve a useful purpose by providing a support for the derricks, which acted as crane jibs for the purpose of loading and discharging the cargo. As shipping progressed with the times, so did the design of the masts. They became shorter, stump masts otherwise known as king or samson posts were placed in all sorts of unorthodox positions, all with the purpose of allowing either additional or stronger derricks to be fitted. Sometimes twin samson posts were connected by a transverse lattice work at their tops so that a pole topmast could be shipped on the fore and aft centre line to carry a steaming light, ships carrying stump masts linked in this manner became known as "Goal-posters".

Today every ship still requires to be fitted with masts if only to carry the steaming lights, aerial and flag halyards. Masts on tankers and other ships that are no longer required to support lifting gear are normally of a very light structure, while the masts of a modern ship fitted with heavy lifting gear are very solid affairs, many of them being built in the shape of bipods or tripods.

Starting with the conventional mast, it would be a strong tubular steel post and originally was stepped on the keelson, later it was stepped on either the main deck or tween deck and housed in a tabernacle (strong vertical casing which would securely anchor the base of the mast), the next step was to dispense with the sides of the tabernacle and at the same time increase the length and width of the top, so that it could house the heels of several derricks, this platform at the base of the mast became known as the mast table. The sides were later put back to make a mast house which could be used to stow the derrick gear.

The mast is further supported above the weather deck by wire rope stays and shrouds and perhaps an additional pair of swifters, all attached to a strong band at the masthead called a "houndsband". At the masthead, above the houndsband, a strong thwartship girder called the "crosstrees" supports the topping lift blocks of the derricks.

The stays are secured to strong points forward of the mast while the shrouds, swifters and backstays which are transverse supports, are all secured to the chain plates, by means of bottle screws or rigging screws.

Although today most ships have a vertical steel ladder attached to the mast, ships can still be seen with ratlines in the rigging. That is to say, a rope ladder made by securing lengths of ratline horizontally between the shrouds at intervals and with sheerpoles. The sheerpole being a horizontal bar fitted with several belaying pins, to which in the days of sail, various ropes could be made fast. The sheerpole passes through the thimbles at the base of the shrouds and swifter and prevents the wires unlaying.

The bottle screws setting tight the stays and shrouds must be kept well greased in order that they may be easily let go, as is sometimes necessary when working cargo. The exposed threads of the bottle screws should have greased rope yarns wound in and be protected by canvas covers called "gaiters", which are sewn in place. Each bottle screw should also have a locking device which will prevent it becoming loose with vibration.

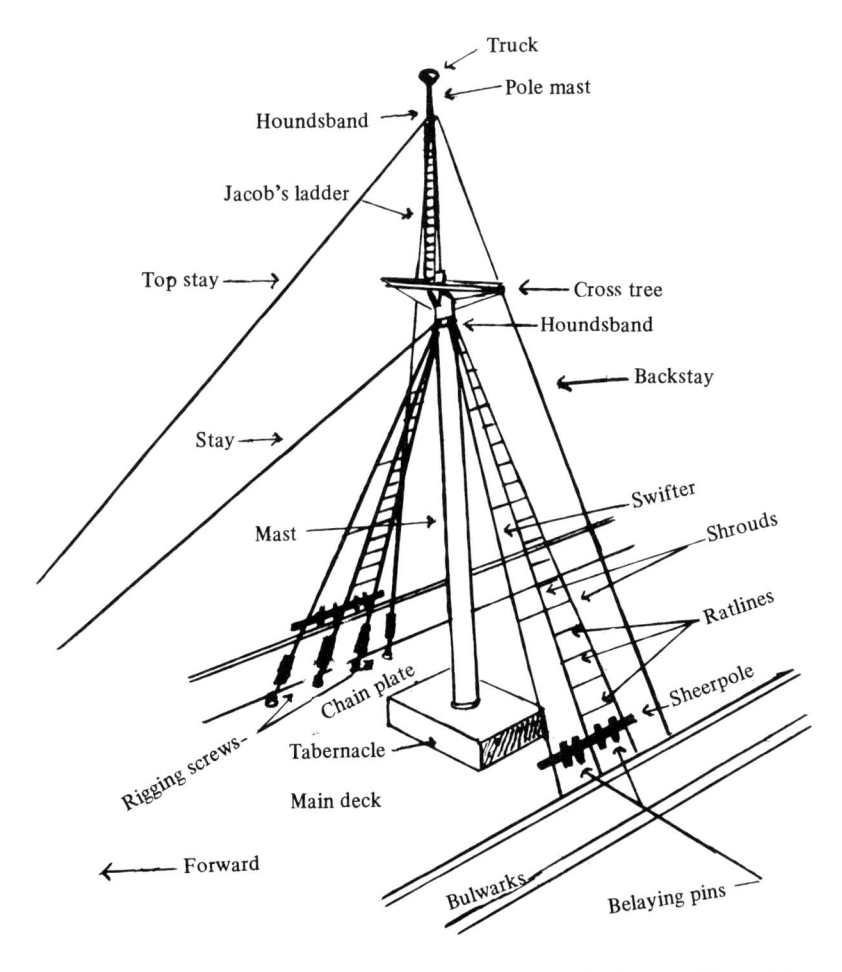

A typical conventional mast showing standing rigging. Stays take the name of the mast, i.e., forestay, mainstay, etc. Note that every fifth ratline is extended to the swifter.
When setting up rigging, set up the stays first followed by the shrouds.

Never carry any tools or gear up or down a mast or hold ladder. Before going up a mast, make one end of a light line fast to your waist and when aloft use the line to haul up any gear you require.

The masts are named from forward to aft Fore, Main, Mizzen, Jigger and Spanker in that order, according to the number of masts stepped in the ship.
Topmasts. Above the mast, stepped at the mast-head will be a light topmast. This is necessary to carry the steaming lights at the required heights, also the aerial. A button on the top, called the truck, prevents rain, etc., entering the mast and a fore and aft sheave, carrying a dummy gantline, will be inserted near the top for use when painting down the mast and when working aloft. A houndsband will be fitted to which the topstay and backstays are secured and when a yard is fitted, the yard lifts. A yard (horizontal spar) is sometimes fitted for the purpose of carrying flag hoists. The yard, if fitted, will be on the fore side of the mast and secured to it by a parrel (collar) at the bunt (centre of yard). The flag halyards will be at the yard arm (end of the yard). A wire going from the houndsband to each yard arm is a yard lift.

Topmasts may be fixed in which case they are known as pole masts, or it may be possible to lower them into the interior of the mast when they are known as telescopic topmasts. Many ships are fitted with telescopic topmasts when building, so that in the event of the ship proceeding up the Manchester Ship Canal, the masts can be conveniently lowered and the ship will be able to pass safely under the bridges.

A telescopic topmast is held in position by a heavy iron pin called a fid and the heel is wedged with wood wedges to prevent movement. A canvas coat is sewn over the wedges to keep the whole watertight.

To lower a telescopic topmast.
1. Get together a good wire runner of sufficient length and with an eye in one end only. (to be used as the heel rope) A snatch block, spike, handy billy, bo'suns chair with a gantline, a 2 lb. (.9kg) hammer, heavy hammer, 5 ton shackle, a length of 2½ inch (20mm) fibre rope, suitable tail block, sack, sharp knife and a safety harness.
2. Obtain power on deck and take the cargo runner off the winch.
3. Send a man aloft. Marry the gantline to the dummy gantline in the sheave at the masthead, reeve the gantline, make fast the bo'suns chair and haul it up to the cross-trees.
4. Make the tail block fast to the cross-trees with the 2½ inch (20mm) gantline rove through.
5. Come up on the topstay and backstays (use the handy billy to hold the weight of the topstay) and slack up the halyards.
6. Remove the canvas coat and wedges from the heel of the topmast, place the mast coat and wedges in the sack and send them down.
7. Using the bo'suns chair have the bare end of the heel rope married to the dummy gantline in the heel of the topmast, pulling it through the hole in the mast and out over the sheave on the other side. The thimble eye in the heel rope must be shackled to a ring bolt or houndsband in such a manner that it will not chafe against the hole in the mast, when the weight of the topmast is upon it.
8. Lead the bare end of the heel rope through a snatch block at the base of the mast, thence to the winch barrel and secure it.
9. Run the heel rope onto the winch barrel and take the weight of the topmast. (do not heave more than is just sufficient to take the weight of the topmast off the fid).

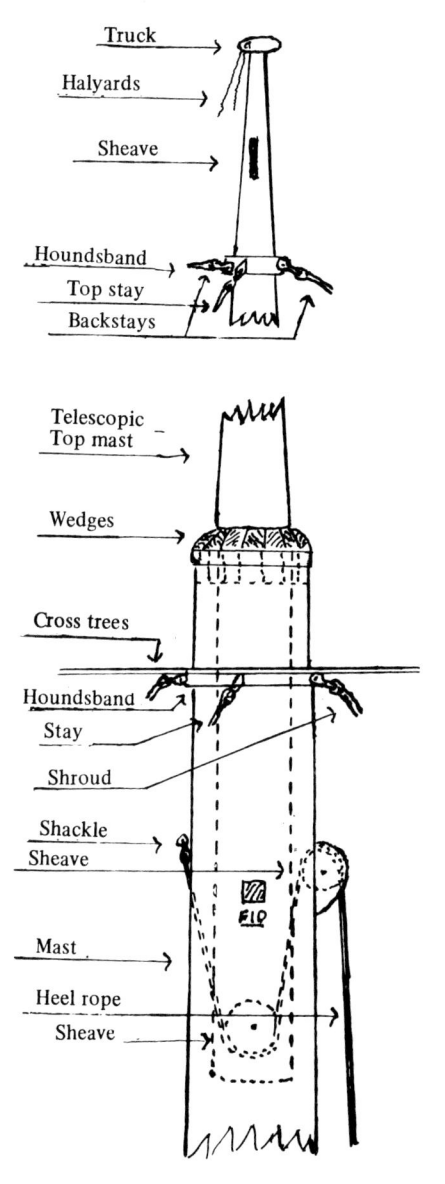

SHOWING THE ARRANGEMENT
OF A TELESCOPIC TOPMAST

10. Using the bo'suns chair, put a line on the fid and then knock it out with the heavy hammer. The fid and hammer should then be lowered to the deck using the gantline rove through the tail block.
11. With the man out of the chair, lower on the winch until the steaming light, yard or any other obstruction is just above the cross-trees. Remove same (the yard may be removed by letting go the bolts in the parrel) and send down anything removed.
12. Continue lowering the topmast until it rests on the mast keepers inside the mast.
13. Take the heel rope off the winch and put the cargo runner back, collect and stow all the gear.

Note. All gear sent aloft to have lanyards attached for securing same when not in use. Any men working aloft to wear safety harnesses throughout the operation, all hands to wear protective helmets and footwear.

To raise the topmast reverse the procedure using the handy billy to set up the topstay.

The wire supports of samson posts and funnels are always termed stays, regardless of whatever direction they may lead.

The masts of tankers often carry the tank venting and possibly pressure valves.

Derricks. In the early days of sail, cargo was loaded and discharged with the aid of a jumper stay. A gin block would be seized to a stay set up between the masts. A rope fall rove through the gin block had a hook on one end for catching cargo. The other end was held by a member of the crew who would jump from the bulwarks or some other place, holding the rope and so lifting the cargo. Hence the name jumper stay.

The next step was to put dolly winches (hand winches) aboard. The fall from the jumper stay could now be led to a dolly winch and a fall from a second gin block secured to the end of a cockbilled yard, would be led from the cargo through the gin block at the yard arm to another dolly winch. It was now possible to transfer cargo from the hold to the quay by using the yard and stay to carry falls that had been joined together at the hook to make a union purchase.

With the advent of steam and steam winches, swivelling booms were attached to the heel of the mast to act as crane jibs for the loading and discharge of cargo. The first of the modern booms or derricks had a single wire topping lift span, led from the derrick head to a block at the mast head and down the side of the mast. A preventer chain was shackled by means of a union plate to the wire topping lift span, and shackled to the deck to hold the derrick at the required height. On some ships a stout manila rope luff tackle was substituted for the chain preventer. The wire span and chain preventer method is still in use on some of the older ships. The next step was to put a wire rope purchase between the derrick head and the mast head, the wire fall of the topping lift was led down to the deck where it could be made fast to a cleat or bitts. From this point it was a simple matter to put the topping lift permanently on to its own winch. Winches which held the derrick at the desired height and had either a pawl and rachet or carpenter's stopper to prevent the derrick from being hove down, were the first, but now it is quite common (especially with heavy lifts) to have the topping lift attached to a powered winch, so that the derrick may be raised or lowered as required with cargo suspended.

To top a derrick with a single wire span and chain preventer.
1. Get together the chain preventer, the two tested shackles for the topping lift, a snatch block, some seizing wire, a spike and if required a preventer guy.
2. Obtain power on deck and take the runner off the winch barrel.
3. Stretch the guys and shackle the cargo end of the runner to the deck. Remove the derrick head lashing.
4. Secure the bull rope to the winch barrel and run it on, leading it through the snatch block.
5. Man the guys, lift the derrick out of the crutch and slip the preventer guy onto the derrick head if required.
6. Top the derrick until the union plate is down to the snatch block. Shackle the chain preventer onto the union plate. Mouse the shackle.
7. Lower the derrick to the required working height and shackle the chain preventer to the ringbolt on deck, taking care to lift the next link above the shackle. Mouse the shackle.
8. Make fast the guys. Take the bull rope off the winch and hauling it hand tight make it fast to a mast cleat to act as a preventer.
9. Secure the runner to the winch barrel and run it on. Return the gear.

To lower the derrick, reverse the process.

To top a derrick with a wire topping lift and no chain preventer.
1. Get together a snatch block, chain or carpenter's stopper, spike, some rope yarns and a guy preventer if required.
2. Obtain power on deck and take the runner off the winch barrel.
3. Stretch the guys and shackle the cargo end of the runner to the deck.
4. Secure the topping lift to the winch barrel and run it on, leading it through the snatch block. Remove the derrick head lashing.
5. Man the guys, lift the derrick out of the crutch and slip the preventer guy onto the derrick head if required.
6. Top the derrick right up.
7. Put a chain stopper onto the downhaul of the topping lift, above the snatch block. Walk back the winch until the weight is on the stopper.
8. Run the topping lift off the winch barrel, take it out of the snatch block and catch a turn on the mast cleat or bitts. Remove the stopper, lower to the required working height, make fast, lash down the turns with rope yarns.
9. Secure the cargo runner to the winch barrel and run it on. Make fast the guys and return the gear.

Note. On some ships the lead for the topping lift is to the winch drum end. In such cases the end of the topping lift MUST be secured to the drum end. Many accidents have occured when only 3 or 4 turns of the topping lift have been taken round the drum end and the derrick topped with a man taking the topping lift as it comes off the drum end. Again on some ships the topping lift remains in a lead block and is made fast to bitts on deck.

To lower:— Man the guys, remove the rope yarn lashing and all but the last two or three turns of the topping lift from the bitts. Slack away on the topping lift. Remove the preventer guy before landing the derrick. Land the derrick, put on the derrick head lashing, make fast the topping lift and make up the guys. Haul the runner tight on the winch.

N.B. A chain stopper should not be secured to the same ring bolt as the lead block.

Derrick with Single Span and Chain Preventer.

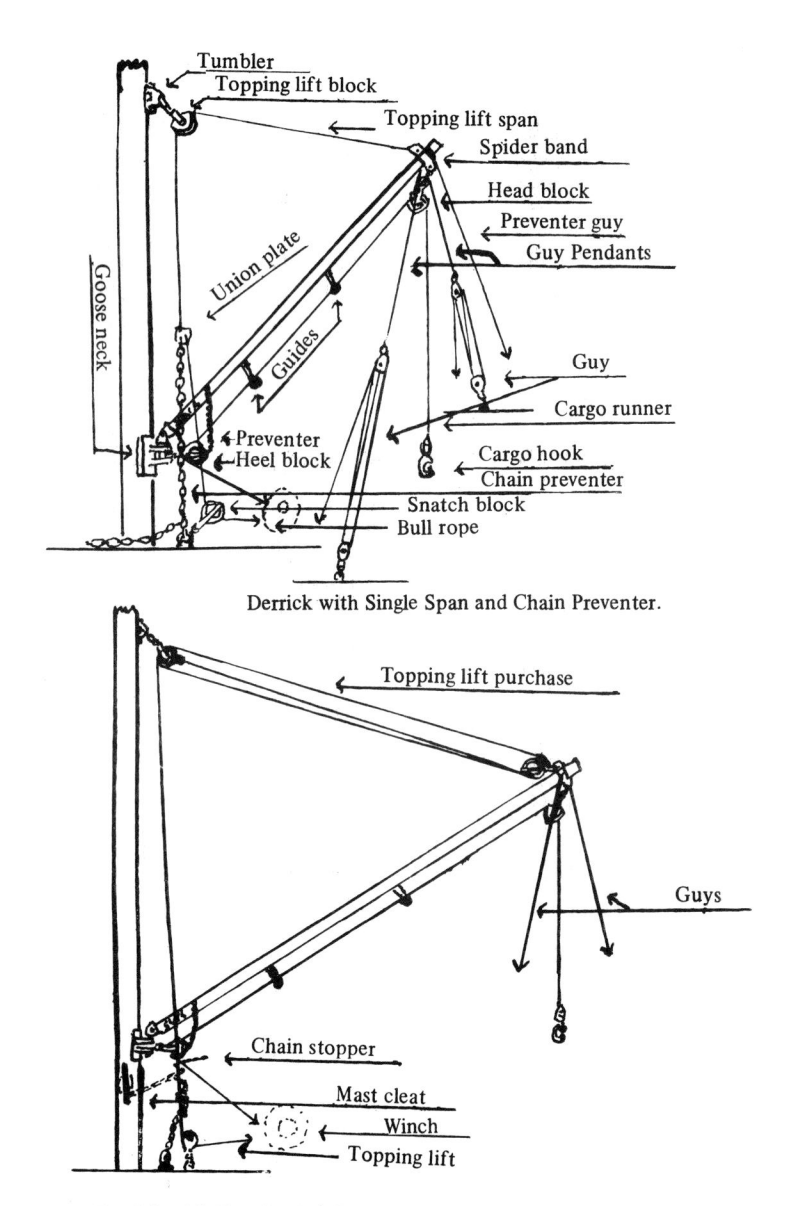

Derrick with Topping Lift Purchase

When lowering on the bitts or cleat:- The topping lift is to be flaked out on deck. A 2nd hand should keep it clear and back-up the lowerer.
Never pay out wire rope from a coil.

Working cargo with conventional derricks.

Cargo may be worked by using a single derrick which is swung while the load is suspended or by using a union purchase, with the derricks rigged in pairs as yard and stay.

When dealing with a swinging derrick, both of the guy hauling parts may be taken to winches, so that the derrick is swung with the aid of power. However, except when dealing with heavy lifts, this is a wasteful method for it requires the use of three winches and additional labour. For light weights the use of one of the winches can be disposed of, by rigging one of the guys with a dead-man. This is achieved by securing a weight (a condemned runner will do) to the hauling part of the guy, so that when the derrick is hauled over by the steam guy (powered guy), the weight is hauled up to the floating block of the dead guy. When the steam guy is released, the weight on the hauling part of the dead guy drops, returning the derrick at the same time. Every care must be taken, when using a deadman, to see that the floating block of the deadman guy is so placed, that whenever the derrick is hauled into place by the deadman (weight), the deadman will not be in a position where it can injure anyone as it descends. An alternative is to have the deadman outboard and dispensing with the steam guy, use the runner of the second derrick as a steam guy, or a reverse of this method. In many ports the stevedores will use their know-how to list the ship just sufficiently to enable them to use a swinging derrick with the minimum trouble.

When lifting heavy weights, a single swinging derrick using two steam guys, must be operated. In addition the weight must not exceed the safe working load of the derrick and the runner must be doubled up. This strengthens the equipment and allows the weight to be lifted more slowly and safely.

To double up:— Lower the derrick and place a single non-toppling tested cargo block on the bight of the runner, between the derrick head block and the eye of the runner. Take the end of the runner round the derrick head, above the spider band and shackle it to its own part. Top the derrick to the required height and hook or shackle the weight to the crown of the non-toppling block. The weight carried by the runner and winch will be halved but the weight on the topping lift will still be the full weight of the load. Where steam winches are employed, the winch employed to lift the weight must be in double gear.

When a pair of derricks are rigged as yard and stay and the runners joined to make a union purchase, care must be taken in setting the heights of the derricks and the positioning of the guys, for if the derricks are at the wrong heights in relation to each other, or the guys are not properly positioned, it is possible to lift one of the derricks as cargo is being worked. This may result in the other derrick being overloaded and hove down thus causing considerable damage and possible injury to persons. No union purchase should be worked unless:—

(a) The union of the runners is made with a proper union purchase treble swivelled cargo hook.
(b) The angle between the runners at the union must not at any time exceed 120°.
(c) The weight lifted on the purchase at any one time should not exceed one third of the safe working load of either derrick.
(d) The arrangement is approved by a competent person before being put into use.
(e) A preventer guy is rigged outboard from each derrick.

On occasion the inboard guys are dispensed with and in their stead, the two derrick heads are linked together with a set length wire rope span. This is known

Reproduced from 'Accidents'

The result of working a union purchase with incorrectly set derricks as a "schooner guy" and when used the two outboard guys should be set up in such a manner that the schooner guy is held taut.

Care of derricks and cargo gear.

Whenever the ship is on a long passage in reasonable weather conditions and in any case at frequent intervals. The derricks should be stripped and all the gear thoroughly overhauled. To send gear down from aloft or to re-rig it, a tail block with a runner long enough to reach the deck should be secured in a suitable position aloft and re-positioned as often as may be necessary, so that the gear may be both hove up and supported in place by the men on deck, using a winch if necessary. A man aloft (especially in a bo'suns chair) should not be required to take the weight of any piece of equipment but merely jog it into position and screw the shackle pin home. Topping lifts, topping lift blocks, shackles, and tumblers are all to be sent down for overhaul. The topping lifts, runners and guy pendants are to be well oiled, all the blocks taken adrift and overhauled, shackle pins and tumblers well greased. The heels of the derricks lifted and the goosenecks greased. When re-rigging, the correct tested shackles are to be returned to their proper positions and well moused. All split pins should be renewed. All guy ropes should be closely examined and renewed wherever signs of excessive wear, chafe or damage are apparent. It is important that the head of a derrick be well secured to the crutch and is unable to slide towards the mast, before any attempt is made to lift the heel, for the purpose of overhauling the gooseneck, otherwise the derrick head may slip out of the crutch. The heel should also be guyed to keep it steady.

When working aloft it is prudent to have a warning notice displayed under, at deck level. "Men working aloft—keep from under". This notice is of particular importance on a passenger ship.

A chain or wire preventer should be fixed between the heel of the derrick and the arse of the heel block, to prevent the block falling when there is no weight on the runner of a topped derrick. In addition the fitting at the crown of the heel block should be duck-billed for the same reason.

There should also be a means of preventing the bight of the runner, between the head and heel blocks, dropping away from a topped derrick and even more particularly from a derrick as it is being topped, when the runner is slack. Normally runner guides are fixed at intervals along the length of the underside of the derrick but in their absence the runner should be rove through the eyes of one or more lizards, secured at suitable positions along the length of the derrick.

Jumbo derricks.

Many ships rigged with conventional derricks are also fitted with a heavy lift or jumbo derrick on the after side of the fore mast. Most are kept shipped, being stowed vertically against the after side of the foremast and are held in place by a collar placed around the derrick head. Some however, are unshipped when not in use and are stowed horizontally on the fore deck beside the bulwarks, being kept in position by strong lashings attached to ring bolts.

The majority of jumbo derricks are rigged to take a safe working load of something in the region of 50 tons although some ships may have a jumbo derrick with a safe working load of as much as 200 tons and the gear on them is accordingly heavy. In view of this, once a jumbo derrick has been rigged, and is likely to be used again within a reasonable period of time, the normal practice is to keep it rigged while the vessel is on passage. Indeed some ships have a canvas cover which can be placed over the topping lift and cargo purchase, to protect them from the weather while the ship is at sea. Nevertheless, the gear must be sent down for overhaul at reasonable intervals.

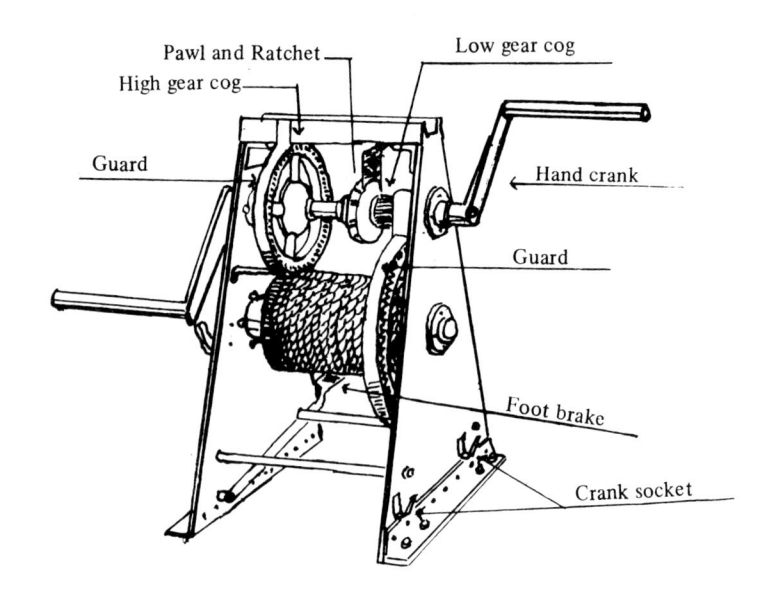

Manual Pawl and Ratchet Winch

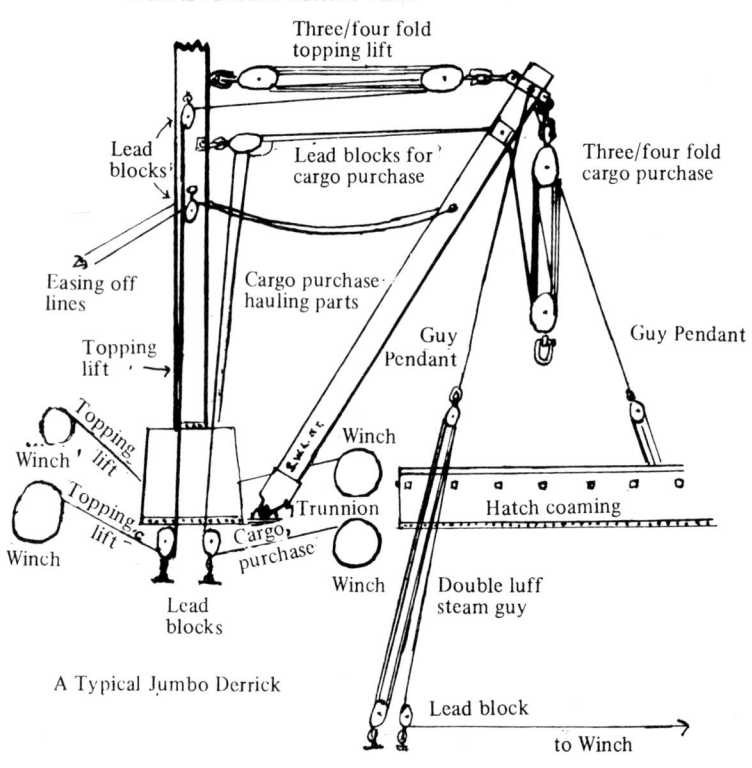

A Typical Jumbo Derrick

When rigging a jumbo derrick, owing to their weight and size, all the blocks should first be hove aloft and secured in their working positions. The wire rope topping lift, cargo purchase and guy falls being rove off afterwards with the help of fibre rope messengers. Similarly, when sending down the gear, all the purchases should be unrove first.

There may be three, four, five or even six fold purchases for both the topping lift and the cargo purchase. They may be rove to advantage, using lead blocks to take the hauling part of each fall from the floating blocks to the masthead, or both ends of each of the falls may be taken over lead sheaves on the derrick head to winches. When both ends of the falls are taken to winches it prevents the floating blocks tipping to one side as a heavy weight is lifted and therefore reduces the strain on both the purchase and the winches.

Steam guy purchases will also be rove to advantage and suitably placed lead blocks are utilised to lead the hauling parts of the guy falls to their respective winches. When lifting heavy weights, it is often necessary to have three separate steam guys, two off shore and one inshore. Sometimes the floating cargo purchase block is fitted with a large steel saucer beneath the crown, to which the guy pendants may be shackled. By transferring the guys from the derrick head to the floating block, the horizontal movement of the load is directly controlled by the guy operators and the derrick head automatically follows. This gives a much better control over the movement of the load than it is possible for guys at the derrick head to exercise. This method however, is only feasible when the weight to be lifted is stowed either on deck, or in the hatchway of the 'tween-deck.

With the jumbo derrick stowed in its collar, the topping lift purchase is slack and the derrick is balanced vertically on its trunnion. In order to lower the derrick, it is necessary to rig a messenger from the derrick head to a lead at the after end of the hatch, then when the collar has been let go, heave lightly on the messenger to get the derrick off balance, it may then be lowered on the easing-off wires until the slack of the topping lift purchase is taken up. The derrick can then be lowered to the required height by the topping lift purchase. Similarly, when stowing the jumbo derrick, it cannot be hove home to its stowed position in the collar by means of the topping lift purchase. When the derrick has been hove as high as the topping lift will take it, the easing-off wires may be used to get it nearer the collar, however it is unlikely that they will be able to heave it home in the collar and it will probably be necessary to rig a fibre rope (to avoid damaging the paintwork) messenger from the derrick head, over the crosstrees and forward to a winch or the windlass drum-end, in order to heave the derrick right home and place the collar in position.

When the jumbo derrick is in use, all four winches at the fore mast have to be utilised. One to take the hauling part of the topping lift, one to take the hauling part of the cargo purchase and two for steam guys. A fifth winch elsewhere may be required for the third steam guy, if one is in use. When both ends of the topping lift and cargo purchases are led to winches, all the steam guys will need to be led elsewhere. Preventer stays are rigged from the fore mast to the break of the fo'castle head, to give additional support to the mast and when steam winches are employed, they must all be put into double gear.

Cranes. In the more modern ships, light 5 and 10 ton derricks and winches, are often replaced by swivelling cranes, so placed that the crane driver is able to see both down the hatch and overside. As the crane driver can not only slew his crane but also raise and lower the jib, the cargo can be more easily placed exactly where it is wanted and the need for signallers is often abolished.

HEAVY LIFT CARGO CRANE

LIGHT LIFT

HEAVY LIFT

HEAVY LIFT

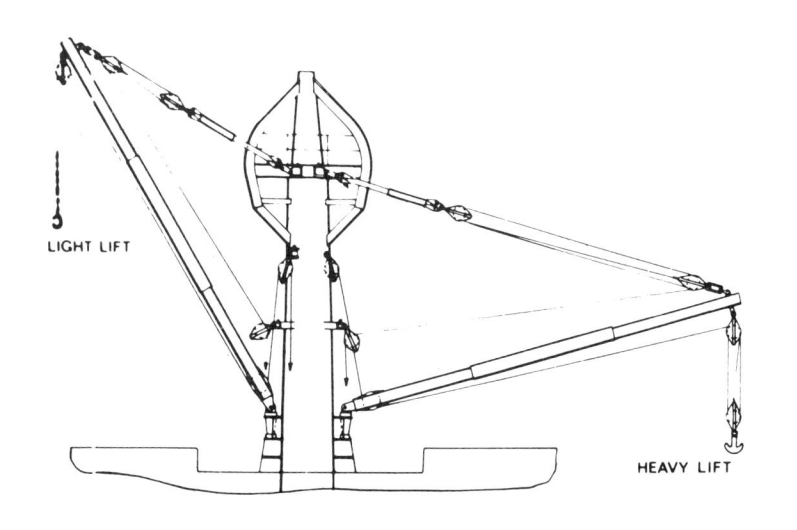

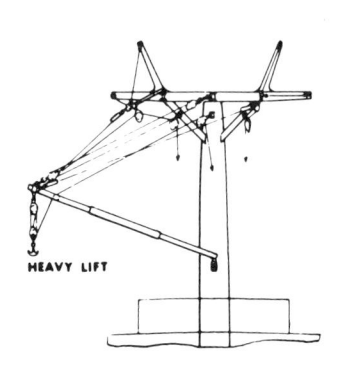

THE HALLEN

SWINGING DERRICK

M.V. 'City of Liverpool'

There is a transverse travelling five ton crane at No.1 hatch. Two ten ton cranes at No.2 hatch and a 33 ton and a 5 ton crane at No.3 hatch. The foremost 33 tone crane derrick serves No.4 hatch together with a 5 ton crane not shown.

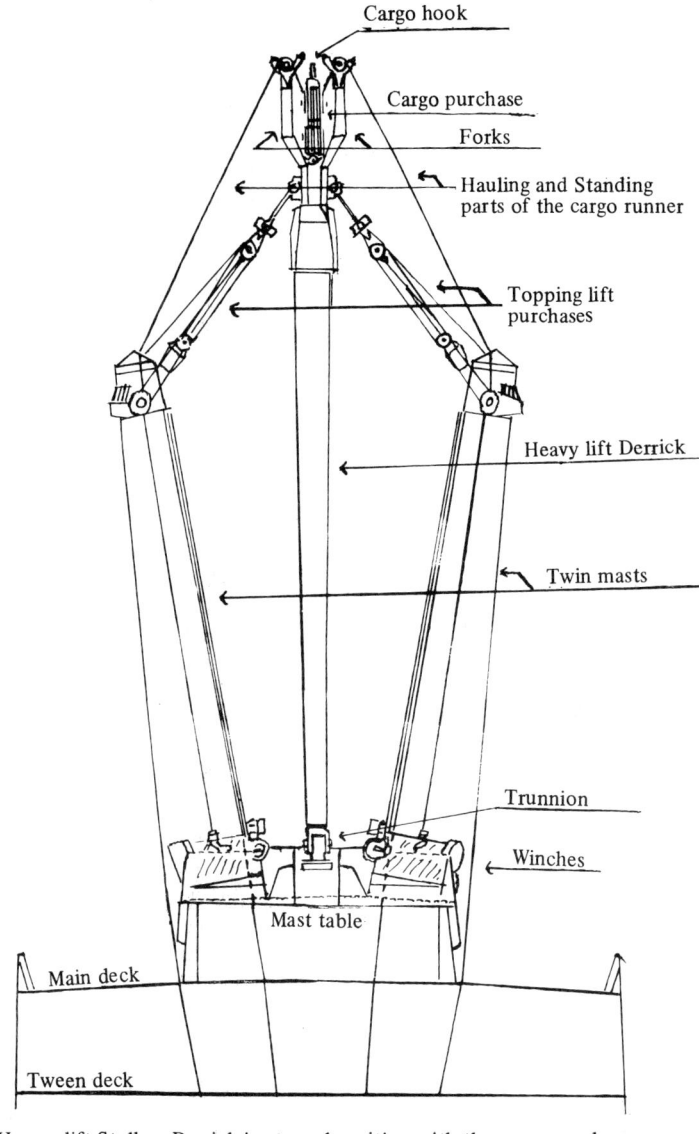

Heavy lift Stulken Derrick in stowed position with the cargo purchase moved "two blocks" and therefore vertical. The Derrick can be operated at either the forward or after hatch simply by being swung over between the twin masts.

For heavier lifts up to about 50 tons various crane-derricks have been patented. Generally speaking they are attached to bi-pod masts and have two topping lifts, while the guys as such are abolished. Three winches with remote control are employed. One for the cargo lifting purchase and two for the topping lifts. The derrick is slewed by means of the topping lifts, hauling on one and slacking away on the other, while limit switches ensure that the derrick cannot be hauled into a dangerous position. Because the controls of these derricks are extremely delicate in action, no attempt should be made to use the controls by anyone who has not received adequate instruction, and is not authorised to drive the winches concerned.

Where ships are required to take heavier lifts, the "Stulken derrick" (of German origin) which can be rigged to lift loads as heavy as 300 tons (depending on the strength of the mast) would appear to be one of the most popular derricks. Erected between twin masts, it has the unique ability of being suitable for use either forward or aft of the mast, being capable of being hove over from forward to aft of the mast (or vice versa) whilst fully rigged, between the masts. It is stowed on passage fully rigged in a near vertical position. The gear is far too heavy and complex to be sent down for overhaul at sea and maintenance is carried out at intervals by shore based staff.

As with crane derricks, no guys are required, the derrick being raised, lowered and slewed by the action of dual topping lifts. Four winches operated by remote control are employed, because both parts of the cargo lift purchase fall are attached to winch barrels, while the remaining two winches each have the hauling part of one of the two topping lift tackles attached to their barrels.

Winches. Today, winches may be powered by steam, electricity or hydraulics. The conventional steam winch is controlled by a steam valve and a reversing lever and has a foot brake attached. Two sets of gears are fitted, either of which may be utilised. The driving shaft has two loose cog wheels running freely upon it, either or both of which can by means of a clutch, be locked to the shaft. One cog is meshed with a large cog-wheel attached to the barrel shaft, the other cog is meshed with an intermediate cog wheel on a secondary shaft. An additional cog on the secondary shaft is also meshed with the large cog wheel attached to the barrel shaft. To obtain single gear (for light loads and speed) the cog on the driving shaft is locked in mesh with the cog wheel on the barrel shaft. To obtain double gear (for heavy loads and slow working) this cog is unlocked and the cog in mesh with the cog wheel on the secondary shaft is locked in position. The second cog on the secondary shaft is then locked in mesh with the large cog wheel on the barrel shaft. The dogs locking the cogs are held in place by levers secured in position with a pin. An additional precaution of binding pieces of wood, cut to the exact length, between the clutches is often employed. When the winch is in double gear, the secondary shaft and barrel shaft can be seen to be rotating in opposite directions.

Before operating a steam winch it is necessary to open all the drain cocks on both the steam and return lines and on the cylinders, to allow any water to drain out, open the valve and allow the winch to tick over slowly. When all the water has been run off, the cocks are closed, oil round and the winch is ready to operate.

Under arctic conditions some ships keep all the steam winches ticking over (out of gear) when not in use, to ensure that they do not freeze up. The writer prefers to set all the winches ticking over (out of gear) and open all the drain cocks on winches, steam and return lines. The main stop valve is then closed and

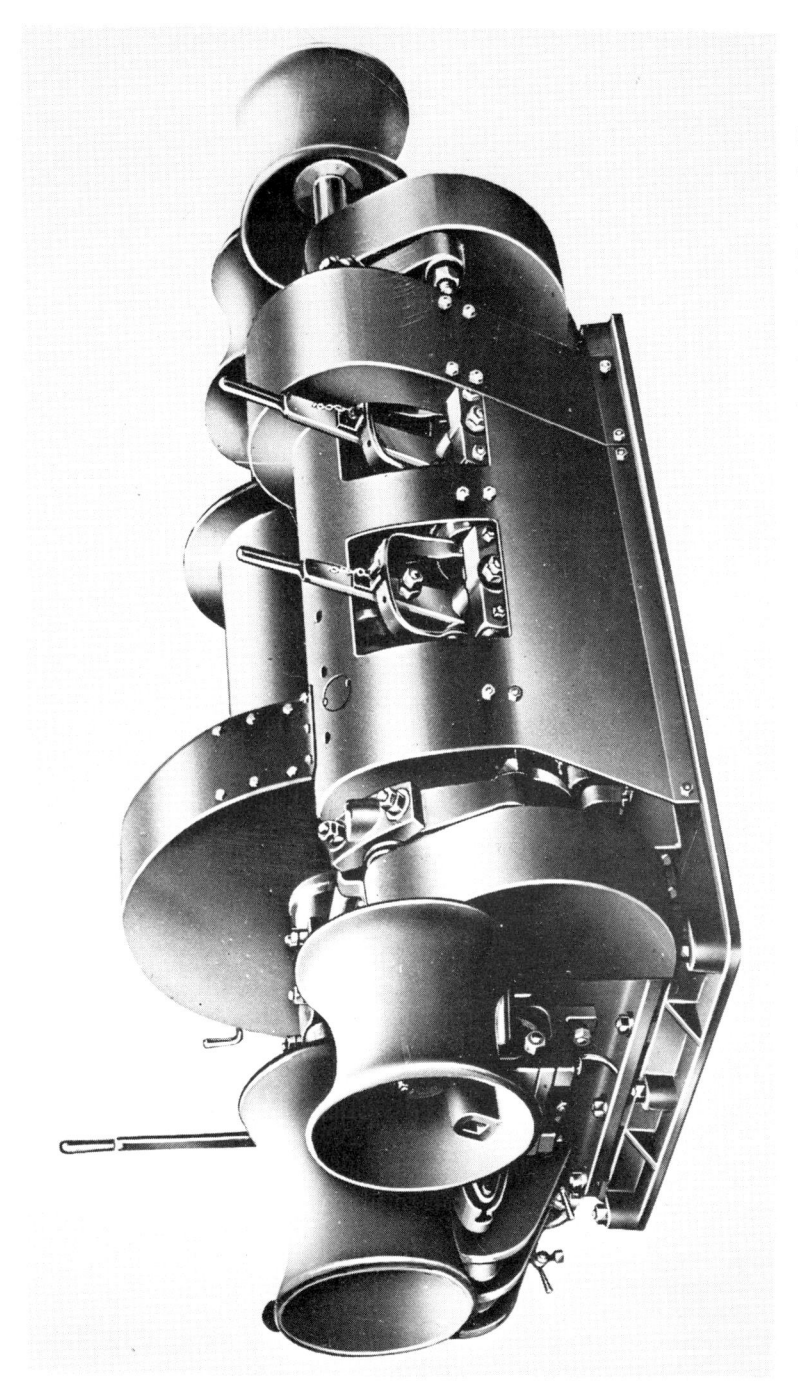

Clarke Chapman - John Thompson Horizontal steam winch standard deck pattern showing levers for selection of single or double gear the last of the steam blows out of the drain cocks, before condensing into water.

When hoisting a load with a steam winch in single gear, if any difficulty is experienced, the winch should be walked back until the load is landed. The winch is then put into double gear and the load re-hoisted.

Electric winches may have the controls placed on the winch or elsewhere on a pedestal or carried by the operator. They do not suffer from freezing problems or require the gearing to be manually changed for hoisting of loads of widely differing weight, however, some do require to be lubricated regularly, usually by means of a grease gun. When overloaded an electric winch will normally automatically cut-out. On some it is still possible to walk back the winch and lower a suspended load by means of power, on others the load has to be lowered on the brake before power can be restored. In the event of a power failure (as against the winch cutting-out) occuring when a load is suspended, an automatic magnetic brake will hold the load. Where an additional foot-brake is supplied, lift the automatic brake while lowering the load by means of the foot-brake. Where remote control with no foot-brake is fitted, a bar or lever inserted in the automatic brake on the winch can be operated to lower the suspended load. In either case the winch control is to be placed in neutral and the power switched off at the main switch box until power is restored.

Hydraulic winches utilise a pump to supply the necessary pressure in the hydraulic system. Like electric winches they have no freezing problems and do not require the gearing to be changed for varying loads. However they do usually require regular lubrication by means of a grease gun. As the system may operate more than one winch, the barrels can usually be put into or taken out of drive, by means of manually operated clutches similar to the clutches operating the gipsies on the windlass and with either hand or foot operated brakes attached.

Self-tensioning winches which may be steam, electric or hydraulic, are fitted to many ships, particularly tankers, for mooring purposes. The mooring rope is secured to the winch barrel and remains there. No stoppering, warping on a drum end, or making fast to bitts is required. The winch can be operated manually in the normal manner, alternatively it can be operated by an automatic self-tensioning device which will remain in active operation for as long as the ship remains tied-up. It is operated by a special pressure-relief valve which automatically operates the winch until a set tension is achieved, when the valve closes and the winch stops. If the tension eases, the winch automatically takes in rope until the tension is restored. If the tension becomes excessive the winch will yeild before the breaking strain of the rope is reached, to restore the original set tension. Automatically allowing for variations in tide and draught.

ACCIDENT PREVENTION.

Never attempt to drive a winch unless you have been properly instructed in the method of driving that particular type of winch.

Never insert your finger when "lining up" bolt holes. Use a spike.

Never catch hold of the floating block of a purchase except by the cheeks. Keep your fingers out of falls, sheaves and hooks.

Never use an untested or under-strength shackle on cargo gear or where it will be required to take a strain.

Never use a rope that has been damaged by kinking or is severly frayed or worn. (Damaged parts should be cut out and the rope spliced).

Never leave a load suspended from a derrick head or other place.

Never walk under suspended or moving loads.

Never secure a runner or fall to a winch barrel with rope yarn. Proper clamps MUST be used.
Never attempt to ride on the end of a runner or crane wire.
Never leave oil or grease lying around. Wipe it up before leaving.
Never use a runner from the derrick head to haul a load out from beneath a deck. Use the runner from the heel block or a bull rope rove through a snatch block.
Never wedge up the pawl on any ratchet winch. Hold it up.
Never attempt to use a winch without someone at the controls.
Never leave a shackle aloft without a mousing.
Never leave the hooks flying when returning empty sling legs by derrick or crane.
Never use power to hoist a man aloft in a bo'suns chair. Do it by hand.
Never leave gear from a telescopic top-mast up aloft while the mast is down, send all the gear down and stow it away until required.
Never attempt to use a winch if the guards have been removed or are in any way unsafe.
Never take tools aloft without a lanyard on them and a bucket to hold them.
Always if possible, make a derrick topping lift fast to bitts with three round turns left handed before figure of eighting and lash the top turns down. Take off the stopper.
Always use shackles in preference to hooks, whenever possible.
Always use two strops when lifting lengthy loads such as pipes, timber, shifting boards, etc. Check the balance of the load before giving the all clear to lift.
Always when hoisting or lowering a weight try to avoid jerks.

Diagram Showing the Component Parts of a Conventional Hatch

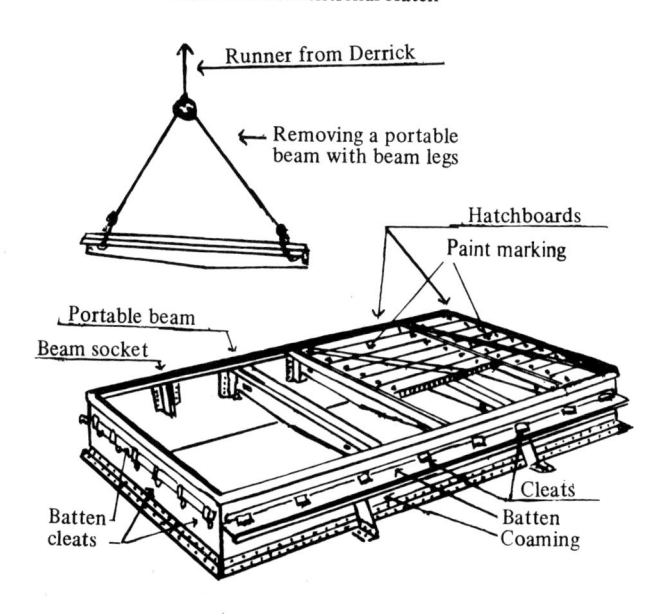

CHAPTER No.11
HATCHES, HOLDS AND DEEP TANKS.

Hatches.

On a conventional ship there will be large openings in the deck called hatches, through which cargo is lowered or lifted as it is loaded into or discharged from the ship's holds. The holds being the very large compartments within the ship in which the cargo is stowed (placed) while being transported from place to place.

Hatches and holds are numbered from forward to aft and since there may sometimes be two hatches to the same hold, hatch numbers do not always correspond with hold numbers.

A conventional hatch on the main deck or any deck above the main deck, will have a steel coaming (vertical plate) at least 2 feet 6 inches (76cm) high and possibly on some ships higher, bounding the hatch on all four sides. 'Tween-deck hatches may be flush with the 'tween-deck, in order that fork-lift trucks may be used for handling cargo in the 'tween-deck, or they will at most have a very low coaming. They will however be provided with portable stanchions and guard-chains for placing around the hatch in lieu of coamings, whenever the hatch is open. Regulations concerning the fitting of these stanchions and chains are contained in The Docks Regulations, 1934 which state:—

Part IV Regulation 37.
(a) If any hatch of a hold accessible to any person employed and exceeding five feet (1.5m) in depth, measured from the level of the deck in which the hatch is situated to the bottom of the hold, is not in use for the passage of goods, coal or other material, or for trimming, and the coamings are less than two feet six inches (76cm) in height, such hatch shall either be fenced to a height of three feet (0.9m) or be securely covered.
Provided that this requirement shall not apply (i) to vessels not exceeding 200 tons net registered tonnage which have only one hatchway, (ii) to any vessel during meal times or other short interruptions of work during the period of employment.
(b) Hatch coverings shall not be used in the construction of deck or cargo stages, or for any other purpose which may expose them to damage.
(c) Hatch coverings shall be replaced on the hatches in the positions indicated by the markings made thereon in persuance of Regulation 14.

It is particularly important for the safety of members of the crew who may work in the holds during the voyage or at a port of destination that 'tween-deck hatchways should be fenced or covered at the outset. Darkness in 'tween-decks after the weather deck hatchways have been covered adds to the danger of any pitfalls that may exist and neglect to ensure the security of 'tween-deck hatchway covers at the time they are replaced may later lead to a fatal accident.

If it becomes necessary to open trimming or other hatchways in the 'tween-decks at sea then the opening should be fenced or covered and well lighted.

Portable transverse beams are fitted between the coamings both to give strength and provide a support for the covering (Except that portable beams are not always fitted when the ship is provided with steel hatch covers of the MacGregor type). These portable beams may be of a lift-out type or it may be possible to slide them to either end of the hatchway, when they are required to be out of the way for the purpose of working cargo.

Portable hatchboards which may be made of wood or metal and fitted to rest on the portable beams, will cover the hatch when it is closed. These hatchboards may be narrow and easily handled, with a handhold provided in the top at each end or they may be large heavy slab hatchboards which have to be removed and replaced by means of legs and a derrick or crane.

All portable beams and hatchboards are required to be clearly marked to show, in which hatch they belong, which deck in that hatch, which section in the hatch, and in the case of portable hatchboards their correct position in that section. It is customary in many ships to mark the hatchboards with differently coloured diagonal painted stripes across each section, so that when being replaced, they are easily placed in their correct positions.

When either portable beams or hatchboards are being replaced, they MUST be returned to their correct positions, the hatchboards are to be kept well closed up together, else the last one will not go in place and considerable difficulty will be experienced getting it properly in position. No hatchboard is to be allowed to overlap either a hatch coaming, a raised flange on a beam, or another hatchboard. When a centre hatchboard cannot be fitted because too much space has been left between the other hatchboards in the section, two hatchboards should be "married". That is to say they are put in place and the adjacent edges are lifted till they meet. The raised part of the two hatchboards is then hit hard a few times with a heavy hammer, so forcing each hatchboard a little outboard and so giving the married hatchboards sufficient space into which to drop flat. Under no circumstances should married hatchboards be forced down by a person jumping on them, most especially when there is an empty hold beneath.

When replacing hatchboards all the outboard hatchboards on one side, starting either from forward or aft, should be positioned first, then the next fleet and so on. This ensures that there is always a hatchboard for the men shipping the hatchboards to stand on. Never stand on a beam, most especially when there is an empty hold beneath. Hatchboards are much easier shipped with the aid of hatch-hooks or chain-hooks than by hand, as is the custom in the majority of coastal vessels, however it does take practice to become adept at handling hatchboards in this manner.

On the main deck and/or upper deck the hatch is then battened down by being covered with three tarpaulins. Tarpaulins are normally marked by having a number of eyelets punched in one corner, equal to the number of the hatch to which they belong. However, corners are often hidden in a folded or "made up" tarpaulin, so that when making up a tarpaulin it is usual to mark it with a number of knots, equivalent to the number of the hatch, in the end of the lashing used to keep the made up tarpaulin secure.

When stretching tarpaulins over a hatch, the seams run thwartships and the overlapping edge of each seam faces aft, so that seas breaking over the hatch are given less chance of splitting the seams open. The oldest tarpaulin is stretched first and tabled, that is to say that the overlapping edges are turned back underneath the tarpaulin so that they do not overhang the edge of the hatch coaming. The newest tarpaulin is stretched next and tucked neatly into the cleats around the coaming. The corners being mitred in the opposite way to what will be the mitring of the corners of the top tarpaulin, in order that the thicknesses of canvas in the corner cleats shall be kept to a minimum. The top tarpaulin is then mitred by taking the overlap at the corner round the corner to the fore-end of the coaming and then tucking the fore-end so that the envelope opening faces outboard. After corners have the overlap brought forward along the outboard side of the coaming and the outboard side tucked over, leaving the envelope opening facing aft. On long voyages it is good practice to stitch up the corners.

Steel hatch battens are then placed in the cleats, outside the tarpaulins, the battens are then secured by having wood wedges hammered well home in the cleats. The wedges are triangular in shape and must be placed in the cleats with the long side of the wedge against the batten. They are hammered home from forward to aft along the sides and from outboard to inboard along the fore and aft ends of the coaming. If wedges are hammered into a cleat with the long side of the wedge against the cleat, they will split along the short grain of the wood and become useless. Note:— On some ships the cleats are so fashioned that the wedges must be hammered home from alternate directions. i.e., one, one way and the next the other way. Wedges should always be hardened up by being driven well home when entering dry weather after wet.

If the ship is loaded, locking bars, which should be served to prevent damage to the tarpaulins, must now be fitted over the tarpaulins or lashings may be used. The purpose of the locking bars, which fit over the hatchboards midway between every two portable beams and between the portable beam and the end hatch coamings, is to keep the hatchboards firmly in position and prevent them floating or falling off, if unfortunately during heavy weather the tarpaulins should become damaged. When heavy seas are sweeping over the hatchways, the eddies formed are broken up by the bars or lashings and tug on the tarpaulins is reduced, thereby minimising the risk of tarpaulins being dragged out of the cleats. They do of course also act as an anti-thief device.

To strip a conventional hatch (open it up).
1. Remove the locking bars or lashings.
2. Knock out all wedges, collect them in a sack and stow them away.
3. Remove the hatch battens. Fold and make up each tarpaulin in turn. Put lashings on them and knot the end of each lashing to correspond with the number of the hatch and stow them away. Replace battens.
4. Remove hatchboards. Centre line fore and aft first and each fleet of hatchboards in turn. Stow the hatchboards one on top of another neatly against the bulwarks or rails.
5. Remove beam bolts from any portable beams it is intended to remove. Plumb the derrick for each beam in turn. Do not go on a beam to adjust the beam legs. Chain on the end of each leg should be passed through a hole in the beam (near the coaming) and shackled back onto the leg. DO NOT USE HOOKS ON PORTABLE BEAMS. Lift the beam and slew the derrick. The beams should be landed flat on the deck, on the off shore side of the ship, resting on pieces of timber, so that when the chain is unshackled it may be drawn easily from underneath the beam. Replace beam bolt in the socket.

When slab hatches or plug hatches are fitted they will have to be lifted from their positions with the aid of a derrick and legs. (Plug hatches may be fitted to an old ship with refrigerated holds).

Beams should never be left upright on the deck, sometimes however, the deck space is limited and the beams have to be left upright. In such cases always wedge the two ends of each beam so that it cannot rock. Hatchboards are not to be used for this purpose.

When rolling beams are fitted, every care must be taken not to tilt or angle them as they are being rolled either forward or aft. A rolling beam that has

Method of Securing a Deep Tank Cargo Hatch Lid

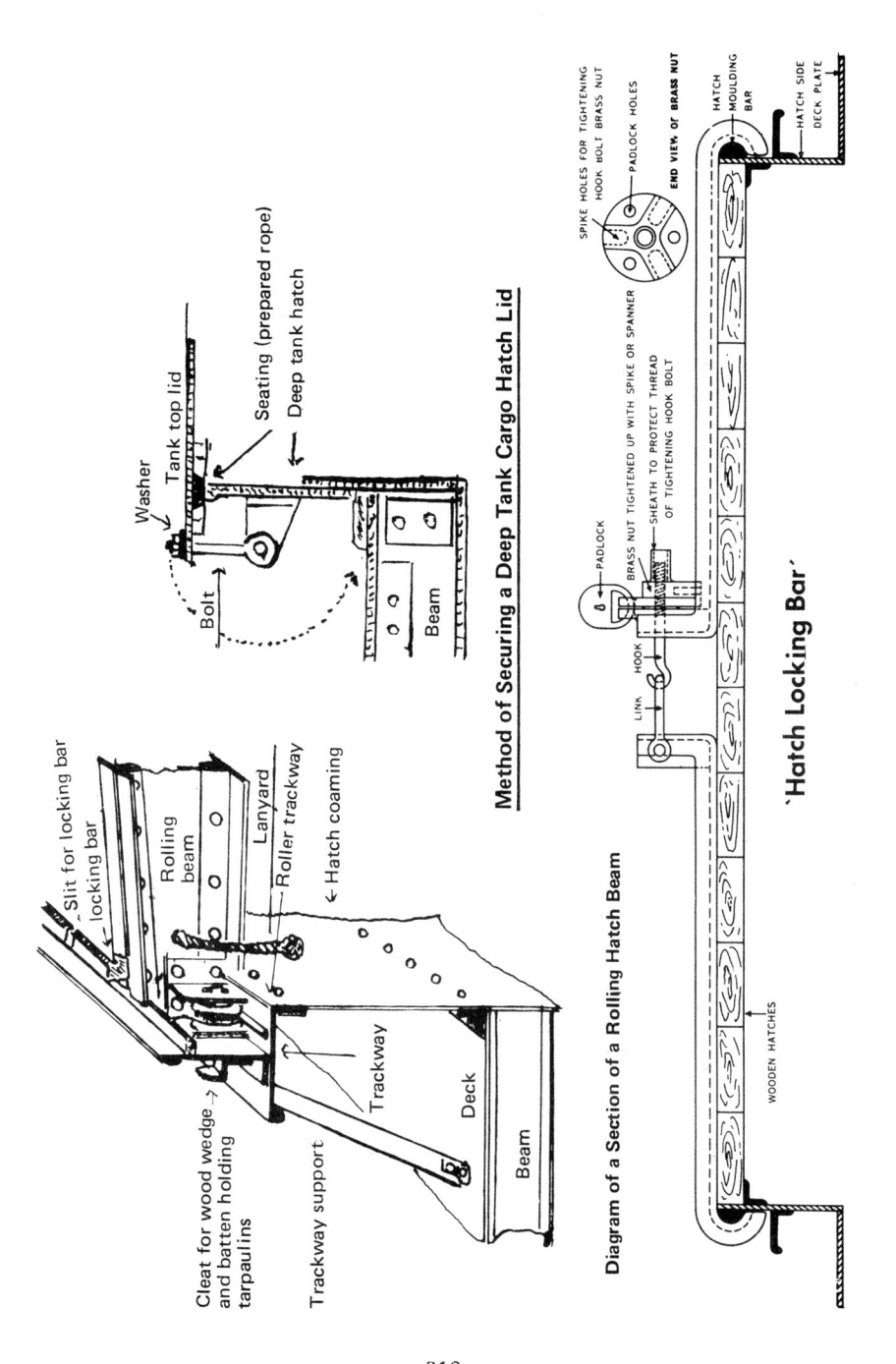

- Washer
- Tank top lid
- Seating (prepared rope)
- Deep tank hatch
- Bolt
- Beam

Diagram of a Section of a Rolling Hatch Beam

- Slit for locking bar
- locking bar
- Rolling beam
- Lanyard
- Roller trackway
- Hatch coaming
- Cleat for wood wedge and batten holding tarpaulins
- Trackway support
- Trackway
- Deck
- Beam

'Hatch Locking Bar'

- SPIKE HOLES FOR TIGHTENING HOOK BOLT BRASS NUT
- PADLOCK HOLES
- END VIEW OF BRASS NUT
- HATCH MOULDING BAR
- HATCH SIDE DECK PLATE
- BRASS NUT TIGHTENED UP WITH SPIKE OR SPANNER
- SHEATH TO PROTECT THREAD OF TIGHTENING HOOK BOLT
- PADLOCK
- HOOK
- LINK
- WOODEN HATCHES

jammed or fallen flat presents quite a problem. Always roll them by hand, never use a bull rope on a winch to pull them.

When some portable beams are left in place while cargo is being worked, they must be bolted or locked in position, to ensure that the cargo runner or hook cannot unship them when cargo is pulled out from underneath.

Hatch tents are sometimes used in port to keep rain out of the hold, when cargo is not being worked. Hook the tent onto the cargo runner and while holding the tent side lashings, hoist the tent over the hatch. Make fast the tent side lashings to ringbolts and hoist the tent until it is sufficiently taut to ensure that the rain will run off.

Many conventional ships are fitted with deep tanks in one or two of the holds for the purpose of enabling the ship to carry a quantity of edible oil. The tanks can of course also be used for ballast when the ship is light. In order that this valuable space is not wasted when it is not required for edible oil, the tanks are normally fitted with large steel lids that can be removed, in order to allow dry cargo to be stowed in the tank space. These deep tank lids which are both watertight and oil tight are secured in place with hinged wing nuts (butterfly nuts) and are seated on the tank coaming on rope or rubber sealing, laid in a channel on the underside of the lid. On some modern ships the lids are hinged and are secured with fixed bolts. The sealing is flax yarn which is laid on the coaming top.

To open a deep tank lid:—
1. Unscrew and turn down all the wing nuts.
2. Plumb the derrick and shackle the beam legs to the ringbolts provided on the tank top lid. Hook on the runner and place suitable wood battens on the deck, where the tank lid can be landed on them.
3. With all hands clear of the lid, lift and slew the derrick to land the lid on the wood battens, taking care not to damage the packing.

To replace the lid, reverse the procedure ensuring that the lid is correctly landed and that the tank coaming is in proper contact with the packing all the way round and will make a watertight joint. Harden up the wing nuts diagonally opposite each other all the way round.

Before closing a deep tank after using it for dry cargo, all dunnage and rubbish must be removed and the scupper covers locked open in case it should next be required to take either a liquid cargo or sea water ballast.

When 'tween-decks are fitted in a ship, the 'tween-deck hatches are not required to be covered with tarpaulins but the beams and hatchboards should be shipped. Some ships will have in addition, four small hatches (one in each corner) known as trimming hatches and which are for use by the dock labour when trimming a loose cargo such as bulk grain. When not in actual use, trimming hatch covers should be kept shipped and when in use should have stanchions and guard chains erected around them. Trimming hatches which are in the 'tween-decks should not be confused with the small booby hatches to be found on the weather deck of some ships. Booby hatches provide access to the holds by means of a ladder and are fitted with hinged, steel watertight lids which are screwed down with wing nuts.

Open tank lids and manholes, when not being worked, must always have the opening fenced or covered when there is no raised hatch coaming.

MacGregor Steel Hatch Covers.

Many ships are now fitted with these covers on the weather deck and sometimes in the 'tween-deck also, especially if a flat surface is required so that fork-lift trucks may be used in the 'tween-decks, or a flat surface is required for roll-on/roll-off cargo. Modern covers can and do vary considerably from ship to ship. On some of the older ships it is necessary to lift the covers manually by means of a steel lever placed in an eccentric wheel (A wheel whose axle is not centrally placed). The covers running on the eccentric wheels have then to be towed off the hatch to their stowage position by means of a bull rope attached to a winch barrel. On more modern ships the covers may still have to be lifted manually with the aid of jacks but can then be stowed or positioned automatically with either an electric motor coupled to chains or hydraulically. The covers may be stowed by being pulled either forward or aft to the stowage position where each section is tipped and stowed vertically. When in one piece they may be pulled sideways, to port or starboard, where they can remain both clear of the hatch and flat. Some are hinged and lifted from one end to a vertical position. In 'tween-decks sliding covers may stack themselves one under the other or be rolled up like a bale of cloth on a bar.

Watertight, oil tight and in the case of refrigerated ships, air tight seals are made between the covers and the coamings by means of rubber packing strips. Where the hatch cover is in sections, steel wedges are driven home over the joints to keep the covers well down and ensure a good watertight joint is made between the covers. The covers themselves are then held firmly in place on the coaming with steel side wedges.

Routine of opening and closing MacGregor rolling hatch covers.
1. Knock out all wedges and side wedges.
2. The check wire should be rigged and secured to adjacent bitts.
3. The haulage wire (bull rope) should then be rigged and any other work requiring a man to climb upon the covers completed.
4. When all men are clear of the covers, each should be raised from it's coaming seating. (Where two sections overlap, always lift the top cover first and if two jacks or levers are in use do not lift both port and starboard sides of the same section together). When replacing securing pins in the bushes of the eccentric wheels, be sure that the pin is so placed that it cannot fall out when the section tips and stows.
5. Two men, one port and the other starboard, should then free the locking pins after first ascertaining that the check wire is secured and all men are clear, including the stowage position at the far end of the hatch.
6. The haulage wire should then open the hatch slowly, the check wire being payed out at the same time, care being taken that the latter does not become taut.
7. When all covers are in the stowed position, the preventer chains should be secured and the haulage wire should NOT be released until such chains are in position.
8. A suitable notice should be affixed in a prominent position at each hatchway. The following wording is recommended by the makers:—
"DO NOT REMOVE HATCH LOCKING PINS UNTIL CHECK WIRE IS FAST AND ALL PERSONS CLEAR OF THE COVERS".

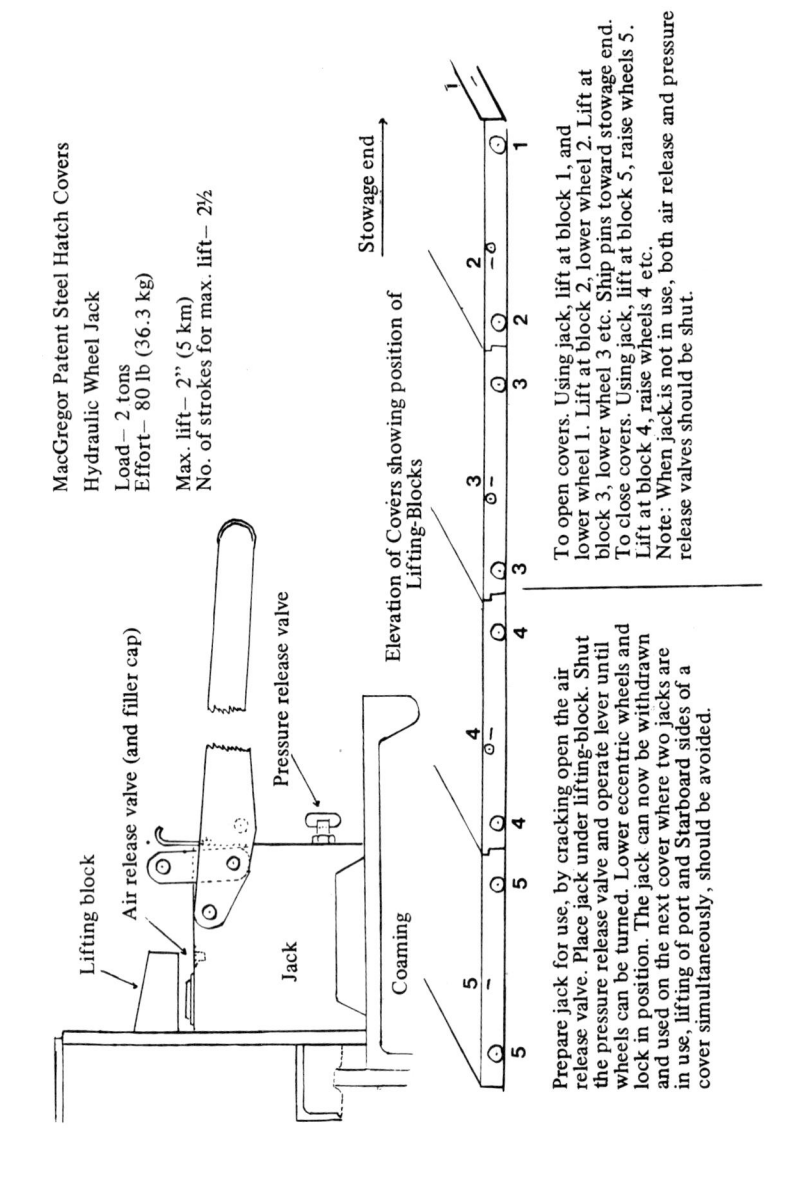

MacGregor Patent Steel Hatch Covers
Hydraulic Wheel Jack
Load – 2 tons
Effort – 80 lb (36.3 kg)
Max. lift – 2" (5 km)
No. of strokes for max. lift – 2½

Lifting block
Air release valve (and filler cap)
Pressure release valve
Jack
Coaming

Elevation of Covers showing position of Lifting-Blocks

Stowage end

Prepare jack for use, by cracking open the air release valve. Place jack under lifting-block. Shut the pressure release valve and operate lever until wheels can be turned. Lower eccentric wheels and lock in position. The jack can now be withdrawn and used on the next cover where two jacks are in use, lifting of port and Starboard sides of a cover simultaneously, should be avoided.

To open covers. Using jack, lift at block 1, and lower wheel 1. Lift at block 2, lower wheel 2. Lift at block 3, lower wheel 3 etc. Ship pins toward stowage end. To close covers. Using jack, lift at block 5, raise wheels 5. Lift at block 4, raise wheels 4 etc.
Note: When jack is not in use, both air release and pressure release valves should be shut.

LOCKING PIN TO BE SHIPPED AND THE CHECK WIRE SECURED BEFORE COMMENCING TO RAISE COVERS.

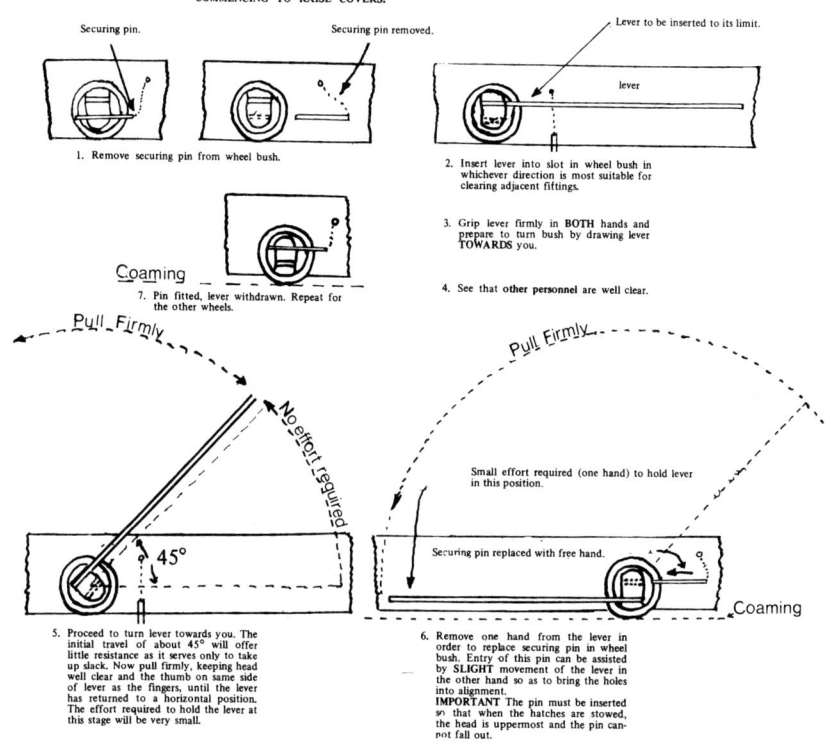

Securing pin.

Securing pin removed.

Lever to be inserted to its limit.

lever

1. Remove securing pin from wheel bush.

2. Insert lever into slot in wheel bush in whichever direction is most suitable for clearing adjacent fittings.

3. Grip lever firmly in BOTH hands and prepare to turn bush by drawing lever TOWARDS you.

Coaming

7. Pin fitted, lever withdrawn. Repeat for the other wheels.

4. See that other personnel are well clear.

Pull Firmly

Pull Firmly

No effort required

45°

Small effort required (one hand) to hold lever in this position.

Securing pin replaced with free hand.

Coaming

5. Proceed to turn lever towards you. The initial travel of about 45° will offer little resistance as it serves only to take up slack. Now pull firmly, keeping head well clear and the thumb on same side of lever as the fingers, until the lever has returned to a horizontal position. The effort required to hold the lever at this stage will be very small.

6. Remove one hand from the lever in order to replace securing pin in wheel bush. Entry of this pin can be assisted by SLIGHT movement of the lever in the other hand so as to bring the holes into alignment.
IMPORTANT The pin must be inserted so that when the hatches are stowed, the head is uppermost and the pin cannot fall out.

IF THESE INSTRUCTIONS AND PRECAUTIONS ARE CAREFULLY FOLLOWED, THE RISK OF ACCIDENT WILL BE AVOIDED.

MACGREGOR PATENT STEEL HATCH COVERS:-

INSTRUCTIONS FOR OPERATION OF ECCENTRIC WHEELS.

TO LOWER COVERS.

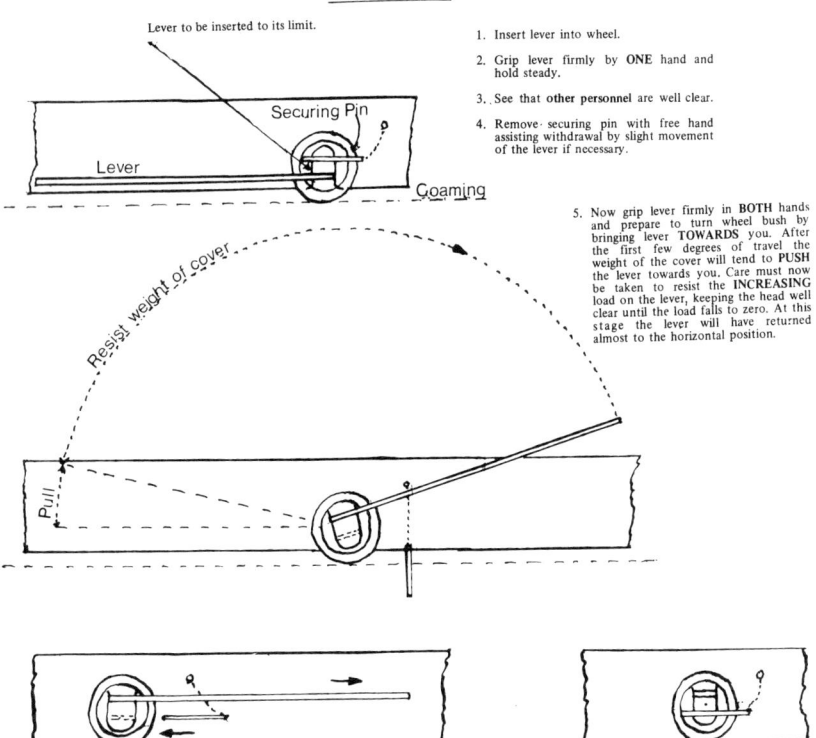

1. Insert lever into wheel.
2. Grip lever firmly by **ONE** hand and hold steady.
3. See that **other** personnel are well clear.
4. Remove securing pin with free hand assisting withdrawal by slight movement of the lever if necessary.
5. Now grip lever firmly in **BOTH** hands and prepare to turn wheel bush by bringing lever **TOWARDS** you. After the first few degrees of travel the weight of the cover will tend to **PUSH** the lever towards you. Care must now be taken to resist the **INCREASING** load on the lever, keeping the head well clear until the load falls to zero. At this stage the lever will have returned almost to the horizontal position.
6. Replace securing pin in wheel bush assisting entry by slight movement of the lever if necessary.
7. Remove lever from wheel bush and repeat for other wheels.

IF THESE INSTRUCTIONS AND PRECAUTIONS ARE CAREFULLY FOLLOWED, THE RISK OF ACCIDENT WILL BE AVOIDED

MACGREGOR PATENT STEEL HATCH COVERS:-
INSTRUCTIONS FOR OPERATION OF ECCENTRIC WHEELS.

Macgregor hatches stowed in the open hatch position.

Macgregor hatches closed.

HOEGH HILL : Deck view showing the arrangement of the forward covers stowing to port and the remainder to starboard.

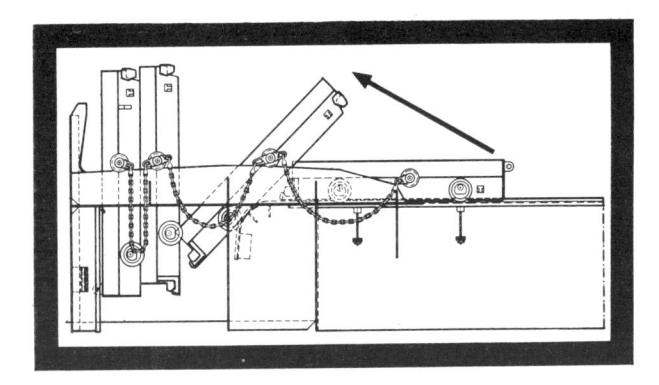

Wire operated "SINGLE PULL" covers

Operation is very simple and is carried out by means of a hauling wire attached to the end panel. During opening, the cover sections are pushed by the end panel. When closing, the sections are pulled by inter-connecting chains or rods. The hauling wire can be led to a winch on deck or winch platform or to a cargo hook.

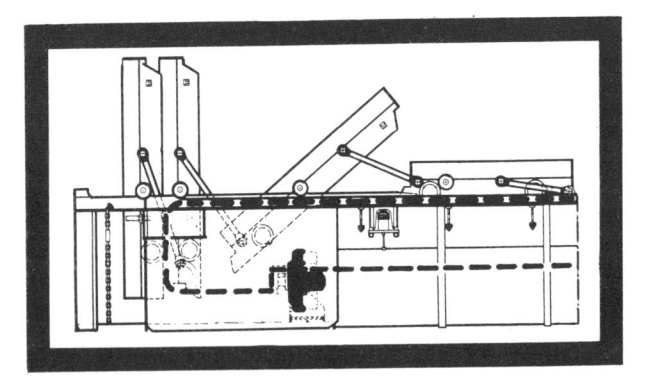

Chain operated "SINGLE PULL" covers
(Long chain model)

Endless chains run the full length of the hatch alongside the longitudinal coamings. The chains which are connected to the leading panel are driven by electric or hydraulic motors, normally situated at mid-length of the transverse coaming. The chains are either led to chain wheels mounted on shaft extensions from the motor or to wheels directly coupled to the motor itself. Opening and closing operations are centrally controlled by lever or push-button. Cover sections are inter-connected by chains or rods.

Locking equipment

Watertightness is ensured irrespective of the type of locking device by special rubber gaskets fitted all round the panels. Locking devices secure the covers in the closed position.

Hand driven wedges
(at inter-panel cross joints)

At each cross joint wedges are fitted on the top plate of the cover section carrying the compression bar. When the covers are closed the wedges are driven to engage on the top plate of the adjacent panel which forces its rubber gasket against the bar.

Automatic transverse cleating
(at inter-panel cross joints)

This comprises two torsion bars fitted at one transverse end of each cover section. Evenly spaced cleating lugs mounted on the bars engage on pressure pads welded to the adjacent cover section which then compress the transverse rubber gasket. Lever arms are fitted at the end of each bar on the outside of the cover side plate. Rotational movement of the bars is produced by the engagement of the lever arms on the longitudinal coaming rail during the closing operation.

Hand operated quick acting cleats
(at longitudinal and transverse coamings)

Incorporated in the head of the cleat is an eccentric which by its 90° rotational movement and self-locking characteristic produces the necessary tension for securing the covers. When the covers are open the cleat is stowed flush with the coaming rail thus enabling the free passage of the cover wheels.

Automatic hydraulic locking
(at longitudinal and transverse coamings)

Sliding bars running the full length and breadth of the hatch are installed below the coaming rail. Locking hooks on the bars engage on spigots mounted on the covers. Each sliding bar is operated by a double acting hydraulic cylinder. The sliding action of the bars produced by the cylinder automatically swings the locking hooks for securing the covers in the closed position. When the covers are open the locking mechanisms are housed flush with the coaming rail. Locking and unlocking operations are centrally controlled at each hatch.

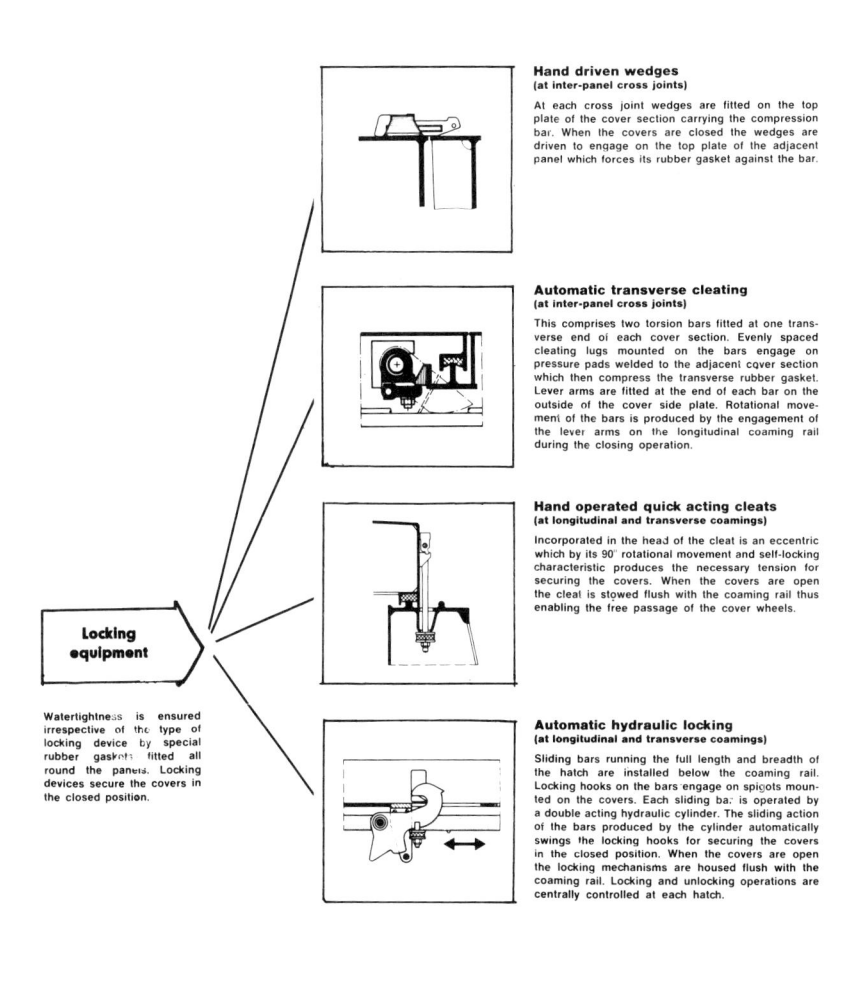

Hand operation

Raising and lowering of small "SINGLE PULL" cover sections, when using eccentric wheels, can be done by means of portable hand-operated hydraulic jacks or hand levers.

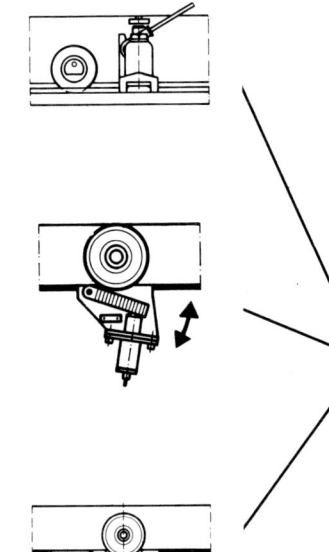

Hydraulic Pot-lift

Remote controlled hydraulic jacks are installed under the longitudinal coaming below each concentric wheel of the "SINGLE PULL" cover sections. These actuate hinged platforms which raise and lower the covers.

Hydraulically operated sliding bars

Sliding bars running the full length of the hatch are installed underneath the longitudinal coaming rails. In way of the cover concentric wheels, lifting pieces slide in small slots cut in the coaming rail. The movement of the sliding bar by hydraulic jacks causes the lifting wedges to engage each wheel, thereby simultaneously raising the cover sections.

Raising and lowering equipment

After unlocking the covers and before opening, the cover sections are raised into the rolling position which at the same time disengages the gasket from the coaming bar thus avoiding damage to the rubber.
After closing, the covers are then lowered ready for locking.

MacGREGOR/ERMANS sliding covers for tweendecks

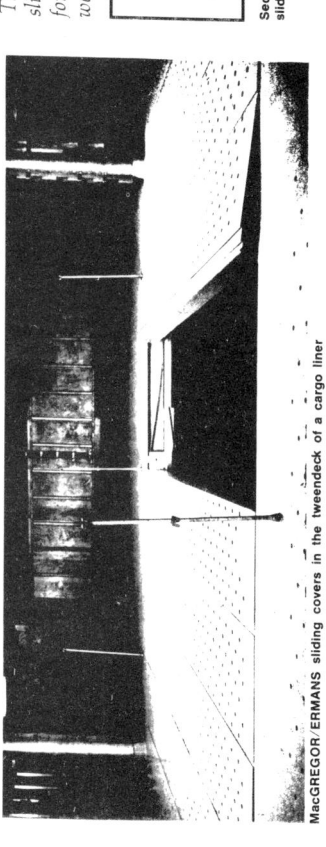

The MacGREGOR/ERMANS sliding cover system is ideally suitable for general cargo liner tweendecks as well as for car decks and extra-decks.

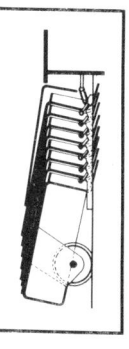

Section through MacGREGOR/ERMANS sliding covers when stowed

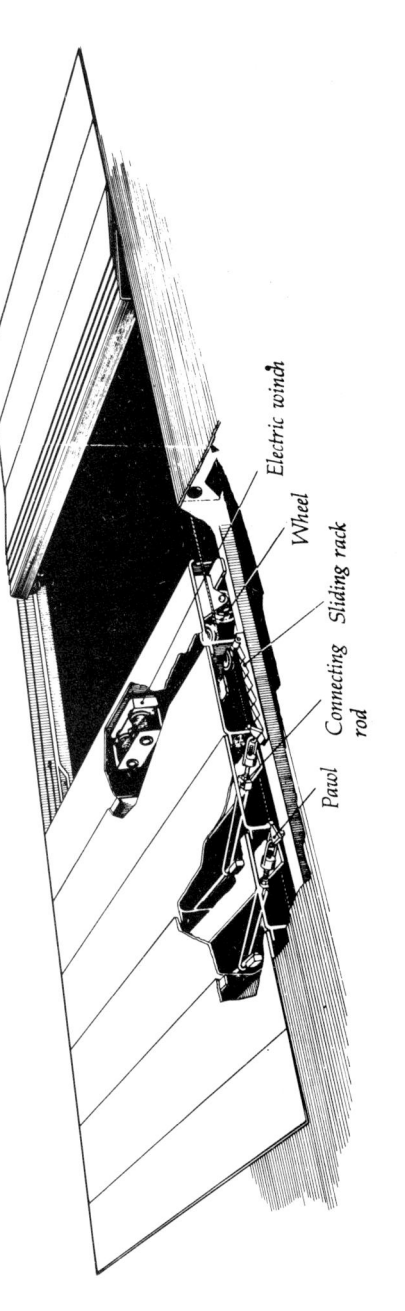

MacGREGOR/ERMANS sliding covers in the tweendeck of a cargo liner

Electric winch
Wheel
Sliding rack
Connecting rod
Pawl

Opening and closing of hatch covers

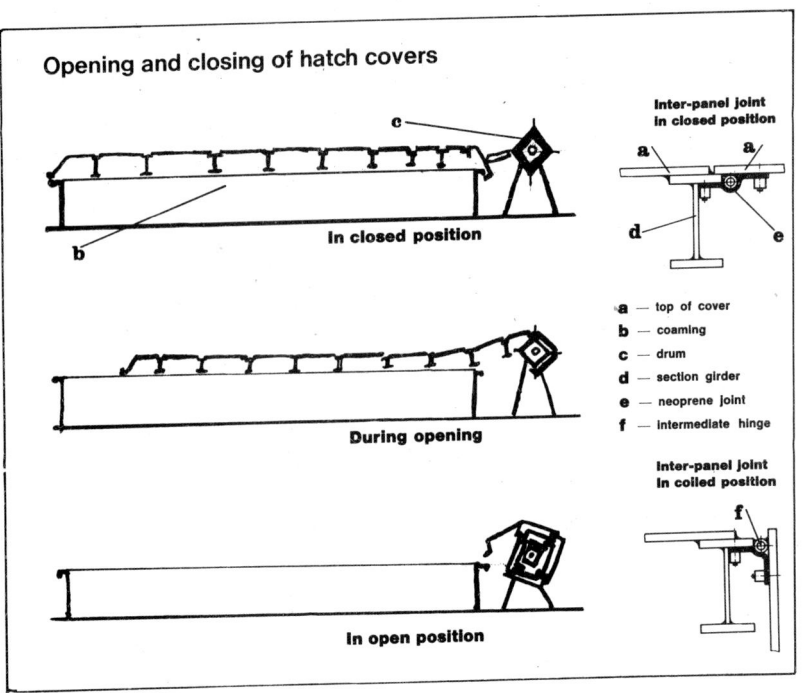

In closed position

During opening

In open position

Inter-panel joint in closed position

- a — top of cover
- b — coaming
- c — drum
- d — section girder
- e — neoprene joint
- f — intermediate hinge

Inter-panel joint in coiled position

The MacGREGOR/ERMANS cover forms a rigid flat topped assembly and is simple to install. All panels are permanently connected by hinges thus avoiding problems of transverse inter-section watertightness and connection. Watertightness between sections is ensured by bolted neoprene bands which are impervious to climatic conditions.

The articulated cover sections have compensating lengths to take into account the coiling action on the powered drum situated at the hatch end. When closing the hatch, the drum rotation is reversed and the panels are pushed along the coaming.

M.V.MORANT – NO.2 HATCH
Showing No.2 Hatch, elevator lid open and winch control platform.

M.V.MORANT – No.2 HATCH LOADING BANANAS.
Showing elevator in position and hatch cover hinges notice the elevators are totally enclosed for cold weather working.

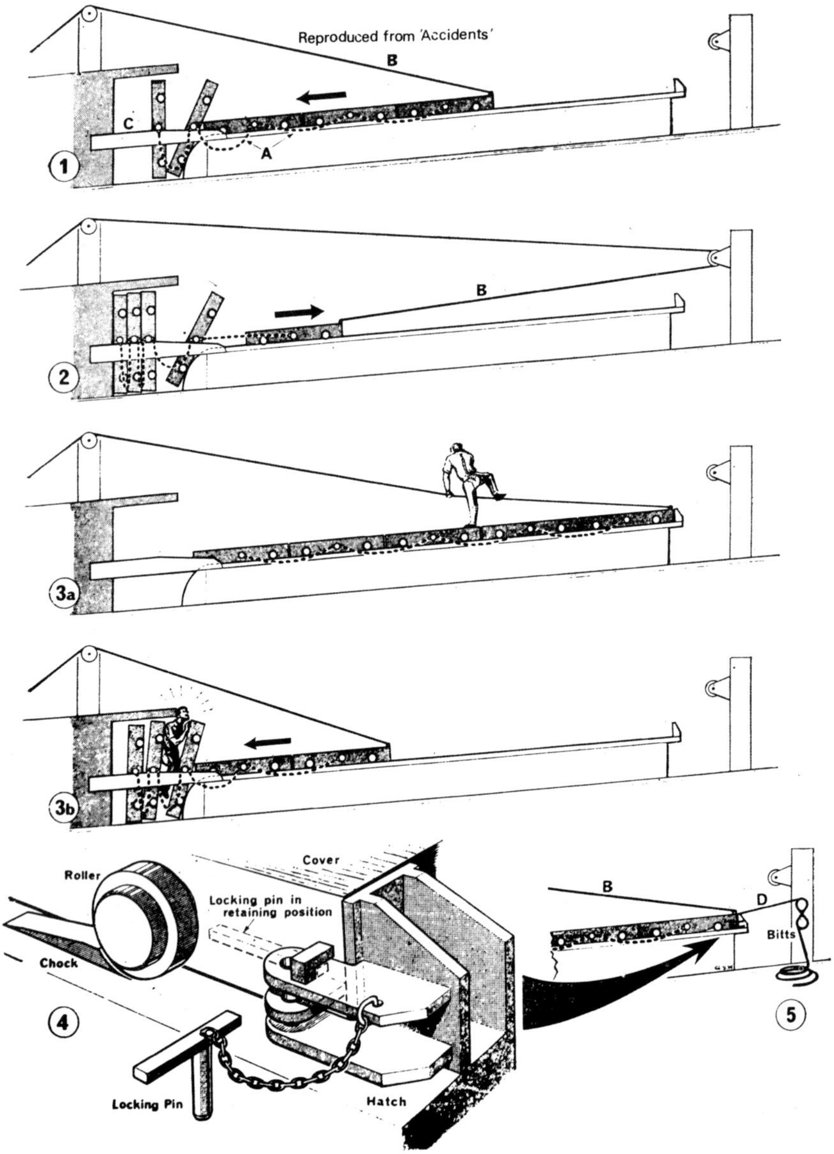
Reproduced from 'Accidents'
B
C
A
1
2
B
3a
3b
Roller
Cover
Locking pin in
retaining position
Chock
4
Locking Pin
Hatch
B
D
Bitts
5

Reproduced from 'Accidents'

In closing the hatch the above procedure should be reversed. It is vitally important that the bull rope be secured to the covers BEFORE the safety chains are let go. Check that all the securing pins are in place, for if a securing pin has fallen out the eccentric wheel will turn as the cover is being pulled into place. Dragging the cover with an eccentric wheel not bearing the weight will in turn ruin the seating, which means a lot of hard work replacing the seating.

Always, whether the hatches are being opened or closed, make sure that the runways for the eccentric wheels are clear. Where the hatches are opened by a pull chain coupled to a motor, do not lean over the hatch as the covers are being pulled on or off. The chains have a nasty habit of jumping unexpectedly and this can cause a nasty injury to anyone standing too near the coaming.

When the rubber seating has to be renewed, remove the old seating by melting the glue, to do this warm the top of the cover with a blow lamp. Cut the new packing slightly longer than is required with vertical joints. Use plenty of solution on both the rubber and the seating bed, take care not to get the solution wet and wait until it has dried. Press the seating into its seat being careful not to stretch the rubber. To help prolong the life of the rubber seating, keep it out of strong direct sunlight and do not let it become contaminated with oil, grease or paint.

Above all, remember that when these hatch covers are lifted, it is quite possible for them to roll on their own, most especially when they are aided by the trim of the ship. Act accordingly, treat them with respect, keep check wires and safety chains in position and KEEP CLEAR.

Holds.

Never go down a hold unless there is someone either with you or standing-by. Make sure that there is plenty of light so that you can both see and be seen. Send down brushes, shovels, etc., on a line, small tools can be sent down in a bucket or canvas bag. NEVER throw anything down the hold. Check the ladder, before going down. If there is wood dunnage below, take a hammer down and use it to knock down any nails you see sticking up. Wear shoes with protected soles and a protective helmet. If wood dunnage is to be stowed in the hold, stack it thwartships against the after bulkhead with fore and aft bearers underneath and take some lashings round the lot so that it will not be thrown about as the ship rolls. Send up timber on the slant with a timber hitch and half hitch. Make mats up into bundles, fold and roll burlap neatly. Send all rubbish up in a bucket, or a canvas bag.

Ensure that the bilges are clean and the strum boxes on the suction pipes have not become blocked with rubbish and silt before closing the limber boards.

When preparing the hold for cargo, dunnage should be laid over the tank top athwartships, so that any water finding its way into the hold has a free run to the bilges. If required, mats are to be placed over all bare iron.

If the ship is to load grain in bulk it may be necessary to erect shifting boards and feeders. Depending upon the ship's stability these may be required to ensure that the grain does not shift. Shifting boards when erected are roughly required to extend about a third of the depth of the compartment from the top and are erected on the centre fore and aft line. They must be stayed from the ship's side with stout wire stays or strengthened with stout wood shores in the same way. If the shifting boards are to be rigged by the crew, as is sometimes required, any man working on the boards as they are being erected is to wear a safety belt. A feeder is a deep box the size of the hatchway placed in the top, to hold extra grain that will feed the hold as the grain settles.

M.E.P. Co. Ltd. Metal Shifting Boards

CONTAINER LASHINGS

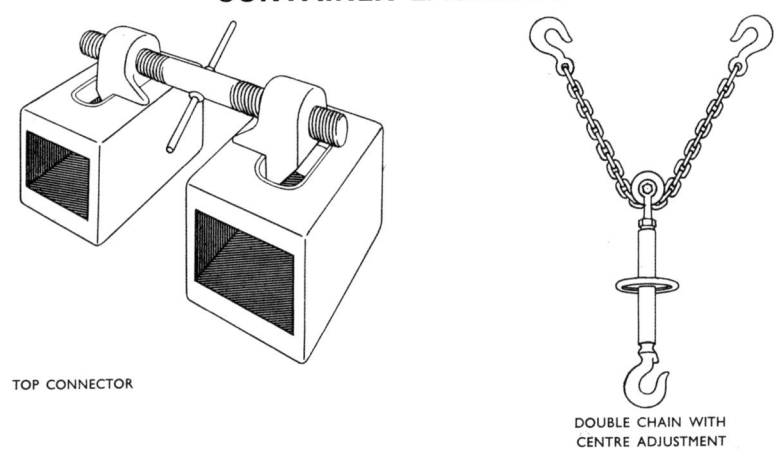

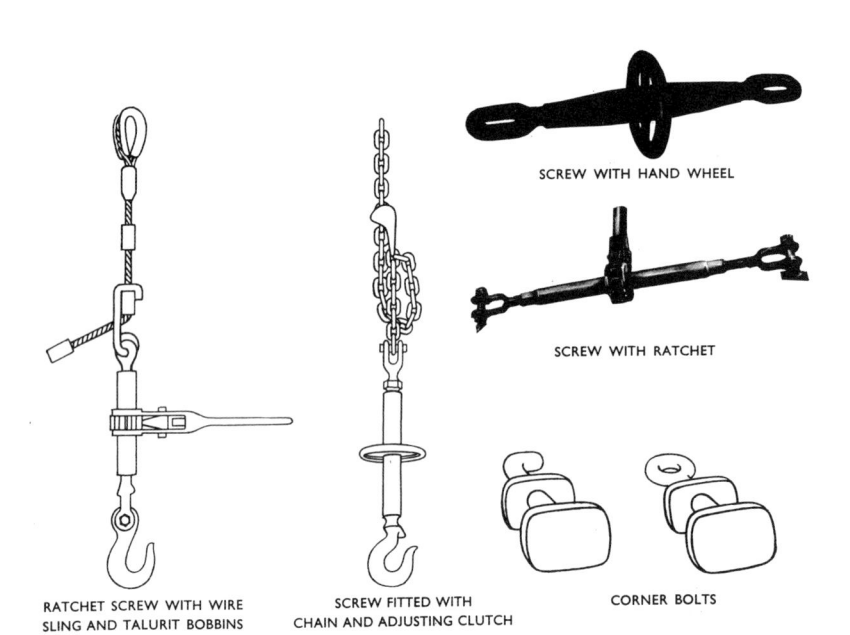

TOP CONNECTOR

DOUBLE CHAIN WITH
CENTRE ADJUSTMENT

SCREW WITH HAND WHEEL

SCREW WITH RATCHET

RATCHET SCREW WITH WIRE
SLING AND TALURIT BOBBINS

SCREW FITTED WITH
CHAIN AND ADJUSTING CLUTCH

CORNER BOLTS

Container deck lashings on an Overseas Containers Ltd., ship

When carrying some types of cargo such as timber, coke or grass, it is usual to carry a deck cargo in addition to the cargo loaded in the holds. Before any deck cargo is loaded, it is essential that the hatches be properly battened down including the fitting of all locking bars. When the cargo has been loaded it will be necessary to have a safe cat-walk over the deck cargo, this cat-walk which is usually constructed with duck-boards is required to have proper stanchions and lifelines erected along its entire length. The derricks may have to be stowed upright, if so, particular attention must be paid to ensure that the lashings are adequate. All lashings on deck cargo are to be tightened daily.

Containers and very heavy items of cargo are also carried as deck cargo. It is again essential that all lashings are tested and tightened daily. The slightest amount of play will allow a heavy object to work, this may eventually lead to the object breaking adrift and so putting the ship in dire peril.

On small ships having rod and chain steering gear, space must be kept clear for the oiling, maintenance and repair of the gear and the cargo so chocked that this space cannot become filled. On all ships carrying deck cargo, sounding pipes and hydrants are to be kept clear.

Ventilation. The holds may be fitted with either a forced or automatically controlled ventilation system but in the majority of conventional ships, natural ventilation is considered sufficient.

When a hold is ventilated by means of cowl ventilators, the lee ventilators should he trimmed to catch the wind while the weather ventilators are turned back to wind in order that they will act as extractors. This provides a good surface ventilation. Cowl ventilators will be provided with canvas covers which can be shipped when for some reason it is desirable to restrict the ventilation. Those fitted on the main deck will in addition be provided with either wood or metal plugs and canvas covers for the ventilator coamings, so that when heavy weather is expected the ventilators may be unshipped and the coamings plugged and made watertight. The best method of unshipping a heavy ventilator is to get three sufficiently stout and long poles rigged as sheer legs (The bottoms spread and the tops lashed together) and lift the ventilator with a double luff tackle. The same method can be used when greasing the ventilator coamings. When cowl ventilators are fitted as column ventilators on the top of samson posts it is usual for them to be fitted with rods and gears so that they can be turned from the deck. However, it is far more likely that the column ventilators will be of the mushroom type and used for extraction purposes only.

When a hold is ventilated by means of general purpose ventilators, they will not require to be trimmed as these ventilators are constructed to both ventilate and extract at the same time, regardless of the wind direction. They may however, need to be unshipped in heavy weather and have the coamings plugged.

Torpedo ventilators are often placed on accommodation bulkheads and provide extraction for the accommodation, alternatively mushroom ventilators may be fixed in the deck-head for the same purpose.

Gooseneck ventilators normally placed on the inboard side of the bulwarks, provide a means of ventilating the double bottom when it is being either emptied or filled. These ventilators are provided with either plugs or caps which must be fitted whenever the vessel is at sea, to prevent any accidental ingress of water into the double bottom. Sometimes an additional plug is fitted in the top for the use of the sounding rod.

In the tropics, windsails are sometimes erected to provide extra ventilation to the engine room, stokehold or the holds. A suitable block is seized to a stay or jumper stay and the windsail hove up on a halyard. Care must be taken to ensure that the canvas trunking is clear of obstructions and able to supply a full volume of air also that it is not chafing anywhere.

ACCIDENT PREVENTION.
Before going under the fo'castle head or into the poop space, ensure that the place is well lit.
When there is a hatch leading to a storeroom or lazerete beneath the fo'castle head or poop or other dimly lit place. Be sure to replace the hatchboards before leaving.
Remember to switch off the storeroom lights before leaving.
Make it a habit NEVER to walk over a hatch except when this is required for the purpose of carrying out your work.
Never close a hatch using only every other hatchboard to support the tarpaulin, anyone crossing the hatch is liable to fall through.
When lowering a light down a hold or storeroom or peak, lower it on a line. Do not rely on the electric cable.
Never walk under working cargo.
When tallying cargo or watching cargo as it is being worked, keep in a safe place. Do not stand under the hatchway or sit or stand either on top of or alongside stacked cargo.
As a member of the crew, your job is to assist in the safe delivery of the cargo. DO NOT BROACH CARGO.
Do not throw lighted matches or cigarette ends or anything else down a ventilator or open hatchway.
When working in a hold with the weather deck covers in place, use plenty of portable lights so that you can see where you are and what you are doing.
Never leave a lighted cluster in the hold, or face down on the deck even when the cluster is switched out. (Someone may switch it on).
In port when battening down hatches temporarily never replace the hatch-boards without the queen beams being in position.
When folding or stretching tarpaulins over a hatch, always look where you are going and pull the tarpaulin behind you.

Merchant Shipping Notice No. M. 292.
HANDLING OF HATCHES LAID FORE AND AFT.
The attention of the Ministry of Transport has again been called to the case of a seaman who was seriously injured by falling through a hatchway while removing hatches.
In the accident mentioned above an A.B. was engaged in removing wood hatch covers from No. 3 hatchway. This hatchway had 4 sections of wood covers laid fore and aft on the usual shifting beams the upper flanges of which totalled 12 ins. wide. Each cover had recessed hand holes and a lifting dog at each end. The man was removing the forward section of covers working from the starboard side towards the centre line, the next section of covers abaft having already been removed. He was standing on the flanges of the exposed shifting beam and in bending down to lift a hatch cover he slipped and fell about 30ft. to the bottom of the hold. The man was wearing rubber boots at the time and the flanges of

the shifting beam may have been rendered slippery by residue from the bulk phosphate cargo which had recently been discharged. No hatch lifting hooks had been provided on the ship.

Accidents of this nature are not unusual, and in order that they may be avoided so far as possible the Ministry recommend that where this type (longitudinal) of hatch is fitted, two iron hooks similar to those used for the handling of chain cable, but about four feet long, should be used for unshipping hatches, the hooks being kept, when not in use, in cleats adjacent to each hatchway. The centre hatch of each section should first be removed and the remainder then turned over and removed in order, working from the centre towards the sides. The hooks may also be used when replacing the hatches, working from the side to amidships.

In this manner, working from the centre towards the side when removing covers, and vice versa when replacing covers, the men will always have the added safety of being able to stand on the hatch covers whilst engaged on the work.

This method has been given practical tests in a ship fitted with hatches laid fore and aft, and the tests showed that not only would the danger be greatly reduced but the hatchways would be more quickly opened up than if the ordinary hand hold methods were followed.

Attention is also drawn to the danger attending the forcing of hatch covers in place by jumping on them. In at least one case in the past the practice has resulted in the death of a seaman.

September 1959. Reprinted December 1966.

Merchant Shipping Notice No. M.409.

USE OF LIFELINES WHEN WORKING IN HATCHWAYS.

The crew of a ship were recently engaged at sea in preparing a feeder in a hatchway to be used when the ship entered port to load a cargo of grain. While one man was adjusting the staging inside the feeder the motion of the ship caused him to let go his hold on the hauling part of the stage rope. The staging collapsed and the man fell to the bottom of the hold. He suffered severe concussion and died almost at once.

The attention of Masters is drawn to the importance of ensuring that lifelines are at all times secured to men when working over empty holds.

July 1957. Reprinted November 1966.

Merchant Shipping Notice No. M.524.

ROLLING HATCH COVERS.

Accidents, some fatal, have occured to men working on single pull steel hatch covers due to the covers "running away" while men have been standing on them whether in the spread or in the stowed position. Men have either been trapped between the sections as they fold up at the end of the hatch or have been caught between the stowage bulkhead or pillar. The danger of such an occurrence is present particularly in ships with excessive sheer of trim.

Seamen and others concerned are, therefore, warned to exercise great care in handling this type of hatch cover. When covers are to be opened up, a check wire should first be rigged and secured. All work requiring men to be on the covers, such as securing the haulage wire and releasing the cross joint fittings, should be completed and the men should stand clear before the covers are put into the rolling position and released.

New hatches of this type are now fitted with a simple safety device comprising a chock on the runway on each side of the hatch and securing pins fitted port and starboard at the end section of the hatch covers. This safety device can also be fitted to existing installations and it is strongly recommended that this should be done at the first opportunity.

August 1967.

Merchant Shipping Notice No. M.992

HATCH COVER SECURING ARRANGEMENTS
THE FITTING OF STEEL LOCKING BARS OR WIRE LASHINGS

(The Load Line Rules 1959 and the Merchant Shipping (Load Line) Rules 1968

Notice to Owners, Masters, Officers and Skippers of Merchant Ships and Sea-going Barges

This Notice supersedes Notice M.666, M.665 and M.513

1. The number of ships lost over the years due to hatch covers not being properly secured has clearly indicated the dangers of proceeding to sea without making use of all the hatch locking bars or wire lashings after the tarpaulins have been battened down.

2. A recent Court of Formal Investigation investigating the foundering of a cargo ship found that the vessel sank due to the failure of the security of one of her hatches and that the vessel was unseaworthy before sailing primarily because the hatch covers were not secured with a full set of locking bars or lashing wires after the tarpaulins had been battened down.

3. The Department of Trade wishes to draw to the attention of Owners, Masters and Skippers that on vessels constructed and equipped to comply with the Load Line Rules, 1959, the steel bars or their equivalent (e.g. approved wire lashings) which are provided for securing each section of hatch covers must be properly fitted in place before the vessel proceeds to sea. (Where hatch covers extend over intermediate supports the steel bars or their equivalent shall be fitted at each end of each section of the covers). The 1959 Rules apply only to those ships whose keels were laid before 21 July 1969.

4. Hatch closing arrangements incorporating portable covers, tarpaulins etc. on vessels built to the Merchant Shipping (Load lines) Rules 1969 impose a freeboard penalty and their use is rare on such vessels. Nevertheless on any vessel where the hatches are closed by means of portable covers the steel bars or their equivalent as required by the Rules should also be fitted in place before the vessel proceeds to sea.

5. The precautions referred to above apply equally whether or not deck cargo is carried on or over the hatches.

6. At sea Masters and Skippers may of course use their discretion as to the occasions on which they may wish to open the hatches as necessary, for

example, for ventilation, inspection, cleaning or preparatory work associated with the next working of cargo.

7. Masters and Skippers are reminded that the stowing of gangways or accommodation ladders on top of portable hatch covers is a dangerous practice which can result in damage to the security of the hatches.

8. Owners, Masters and Skippers are also reminded that proceeding to sea with improperly fitting hatch covers, defective cleats, wedges, battens or tarpaulins constitutes non-compliance with the "Conditions of Assignment", and such is an offence under Section 3 of the Load Lines Act 1967.

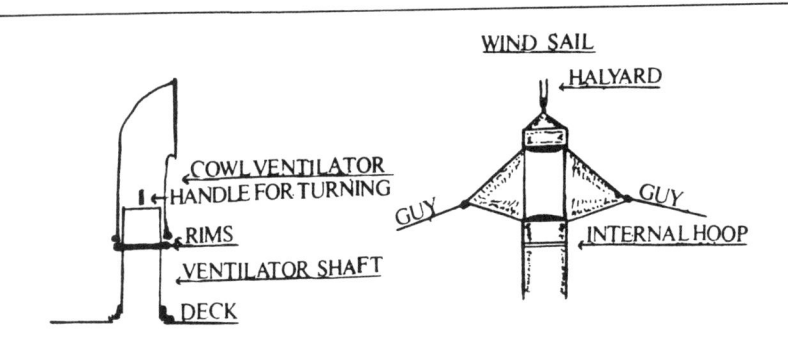

WIND SAIL
HALYARD
COWL VENTILATOR
HANDLE FOR TURNING
GUY
GUY
RIMS
INTERNAL HOOP
VENTILATOR SHAFT
DECK

CHAPTER 12.

MAINTENANCE.

Painting.

Apart from the efficient lubrication of all moving parts, probably the most important part of ship maintenance is the preservation of the hull and its component parts. This can only be achieved by eliminating, or since this is normally too costly, preventing or reducing corrosion. To prevent metal corroding it is necessary to coat the metal with a preparation that will exclude the atmosphere. Normally paint is used for this purpose, some fittings may be galvanised, it is nevertheless usual to paint over the galvanising. Do not use a lead based primer or paint on galvanising or aluminium.

Today there are many different kinds of paint and it is not feasible to expect the average seaman to have any more than a general knowledge of the subject. The days when a Bo'sun mixed his own paints have long since disappeared.

Starting with an old ship. Before painting, all superstructure paintwork must be thoroughly cleaned by washing with sugi (soda) or other solution and all dirt, oil, grease, etc., removed. The paintwork must then be rinsed with clean fresh water and all trace of the soda or other cleaning agent removed. Loose paint and scale is lifted with a scraper, heavy rust will usually loosen when hit with a flat headed hammer. Blunt chipping hammers can be used to remove very thick coatings of loose old paint but sharp chipping hammers should not be used on either rust or paint, they only mark the metal and make matters worse. (Neither chipping hammers or flat headed hammers or scrapers are to be used when there is any possibility of combustible gas or vapour being present). Finally wire brush all bare metal to remove the last of the scale and coat with two coats of a suitable primer followed by a coat of undercoating. The whole of the surface to be painted is then given a coat of undercoating followed by a top coat of gloss or enamel. The surface should always be clean and dry and each coat of conventional paint should be given at least 24 hours in which to dry before the next coat is applied. However, many modern paints will dry in a few hours, some can even be applied to a damp surface, others do not require an undercoat but before attempting anything unorthodox it is necessary to be quite sure that the paint is suited to the use to which it is proposed to put it.

It is unusual in practice (even if it is wrong) to wash topside, boot-topping, hold, tank and bituminous paints before re-painting, neither is it usual to apply undercoating before applying these paints though any bare metal should be given a coating of primer before being painted. Bituminous paints may be applied without previous application of a primer beforehand. However, in every case any grease, oil, rust, scale and salt must be removed before painting.

A new ship may well have been sandblasted and primed by the builders who have then applied one of the more modern paints. In such cases it is hoped corrosion will not rear its ugly head for many years. Maintenance is limited to cleaning the paintwork and touching-up. Future coatings of paint will probably be applied by shore based staff at intervals of anything up to five years, whereas, when using conventional methods it is generally considered necessary to paint the exterior plating two or three times a year, owing to the effects of salt and varying climatic conditions.

Various types of paint.

Blast primers	Applied by the shipbuilder to recently blast-cleaned material, during building.
Primers.	Applied to seal an unpainted surface.
Undercoats.	Applied on top of primers to provide a base colour and a good key for the top coat.
Top coats.	Applied over undercoating to provide a hard wearing surface.
Fire-Retardant.	Has low flame spread characteristics and is for use on exposed surfaces in corridors or stairways and the concealed surfaces of bulkheads, stairways and wood grounds etc. in accommodation and service spaces. Supplied in undercoats and top coats.
Heat-Resistant.	For use on hot surfaces.
Cement wash.	Conventional coating for domestic fresh water tanks, peaks and bilge spaces.
Anti-corrosive	This term is usually reserved to describe the primer on the ships' bottom.
Anti-fouling.	poisonous to marine life, used as a top coat for the ships' bottom.
Boot-topping.	Paint used on the exterior of the hull area that is out of water when the ship is light and under water when the ship is loaded.
Top-side.	Paint used on the exterior of the hull above the water line when the ship is loaded.
Red lead.	An oil based primer. (unsuitable for aluminium).
Zinc-Chromate.	An oil based primer suitable for aluminium.
Flat paint.	An oil based undercoating.
Gloss or Enamel	Top coat for exterior surfaces.
Semi-gloss	Top coat for interior surfaces.
Zinc.	A primer paint.
Bituminous paint.	Suitable as both primer and top coat, for use in peaks, tanks, holds, bilges, and similar spaces. Applied in confined spaces as the ship is building. Requires adequate ventilation when being applied. Unsuitable for use in domestic water tanks, food storage spaces, on surfaces exposed to sunlight, and for tanks that may contain oil, petrol or other solvent. Suitable for use as an anti-corrosive primer on the ship's bottom if aluminium is added.
Chlorinated rubber.	Produced as both primers and top coats, for use on exterior surfaces. Quick drying.
Aluminium paint.	Suitable for use in holds where it reflects light. May also be used as a heat resistant paint. It is sometimes possible to cover a bituminous paint by using aluminium paint as a sealer.
Vinyl paints.	Supplied as primers, undercoats and top coats, for use as boot-topping, top-side and exterior superstructure coatings. Quick drying.

Polyurethane	A two pack paint, the base and hardener being supplied in separate packs for mixing at the time of use. Maximum pot life at 65°F (18°C) 24 hours. Unsuitable for use in cold temperatures. Quick drying. When applied in confined spaces, adequate ventilation is essential.
Epoxy paints.	A two pack paint, the base and hardener being supplied in separate packs for mixing at the time of use. The hardener must be mixed into the base. Very short pot life, ½-hour upwards dependant on temperature and paint, once the pot paint shows appreciable thickening it must be discarded. Quick drying. Supplied as both primer and top-coat. Suitable almost anywhere but the time available for applying additional coats is limited as each coat must be applied before the last has properly dried out. Originally introduced as a coating for the cargo tanks of oil and chemical tankers as it is unaffected by most chemicals.
Polyurethane Epoxy and some other paints.	CAUTION. Adequate ventilation must be provided, care must be taken to avoid contact with the skin and contamination of the eyes with fumes of base paints and hardeners both separately and when mixed. The use of gloves and a barrier cream for the hands and face; protection for head and eyes with helmet or goggles, and the use of protective clothing are recommended. If the skin is splashed, the paint should be removed immediately with a suitable cleansing cream, the splashed area should then be thoroughly washed with soap and water. If the eyes are affected, copious quantities of fresh water should be used.
High build.	Many of these paints can be used as high build which simply means they are put on thickly.
Creosote.	Distilled from coal tar used as a wood preservative.
Stain.	Undercoating and sealer for wood it is intended to varnish, or may be left as a top coat.
Varnish.	A clear or opaque top coat for interior and exterior woodwork or to protect decorative paintwork. For good results must be applied on a hot dry day. Moisture will cause the surface to bloom. (discolour).
Non-slip Paint	For use on weather decks, a two pack paint with aggregate added.
Wet Surface Pretreatment.	Some epoxy paints are suitable for application on wet surfaces, provided they have been freshly blasted. Rainwater, sea water and condensation in close contact with a freshly blasted deck surface can be removed by the application by spray, watering can or brush of a wet surface pretreatment. The excess water and pretreatment solution must be removed by brush or squeegee within five minutes and the epoxy paint

applied immediately. Damp surfaces do not require wet surface pretreatment and in any case over application of the solution endangers the adhesion of the paint.

Paint application.
Generally speaking all the conventional paints and single pack new paints require to be thoroughly stirred before use unless the paint contains a jell. (Paints containing a jell (non-drip) are not normally suitable for marine use). Two pack paints need mixing. Some paints are unsuitable as top coats on primers or undercoats of a different type of paint. Very often special thinners and separate special brush cleaners have to be used, moreover the thinners and brush cleaners are very often not inter-changeable. Fire Retardant paints are sometimes mixed with water or a special fluid. The net result is that ALL INSTRUCTIONS MUST BE CAREFULLY READ AND CLOSELY ADHERED TO. The proper thinners, cleaners and solvents for the paint in question must be used in the proper proportions and the paint may only be put to the uses stated in the instructions, it must not be used on top of either unsuitable other paintwork or an unsuitable surface, furthermore it must be applied by whatever manner is described in the instructions.

Paints may be applied by means of brush, roller, conventional spray or airless spray, many paints can be applied by any of these methods but some can only be properly applied by one method.

Paint brushes.

Pencil.	Small round brush, suitable for figures, letters, etc.
Sash Tool	Small round brush somewhat larger than a pencil brush.
Fitch.	Small flat brush, suitable for figures, letters, etc.
Flat.	In various sizes from ½" (1.26cm) upwards suitable for work anywhere that a roller or spray is unsuitable or when the paint is unsuited to other means of application.
Round.	Long bristled, it helps when the brush is new if the bristles are bound at the brush with light canvas or other material which is then again seized on half way down the length of the bristles and turned over. This shortens the length of the bristles until they become worn down. Generally used for top-side, boot-topping, hold and bituminous paints when these are applied by brush.
Man-help.	A long-handled round brush giving extra reach, sometimes referred to as a "Turks-head" (as distinct from the ornamental knot bearing the same name).

When using a paint brush, always wipe surplus paint off the bristles onto the inner edge of the paint pot and keep the brush strokes in the same direction. Excess paint on a vertical surface, particularly conventional gloss and enamel, will run, with the result that the finished job will be anything but satisfactory. After using a conventional paint clean the brushes in turpentine, paraffin or hot soapy water. When using many of the modern paints a stated fluid must be used for cleaning brushes. After cleaning, hang the brushes up or lay them flat, do not allow any brush to stand indefinately in the cleaning agent or in water, because this will ruin the bristles.

A new brush that is inclined to shed its bristles may be tightened by holding it with the bristles uppermost and wetting the wood at the base of the bristles. After a few hours the wood will swell and tighten its hold on the bristles.

Rollers.

Rollers are made in various lengths to suit differing conditions, the paint is kept in a special tray having an inclined or sloping base. The roller is rolled into the paint and then rolled over the surface to be painted. Large areas can be painted very quickly using this method which is suitable for most but not all paints.

Conventional spray.

The paint in a container is drawn through a nozzle by a current of pressurised air. This is a very fast method of painting but is unsuitable for use in confined spaces. Goggles should always be worn when using a spray gun and vapour should be quickly dispersed by good ventilation when the gun is used under cover. Suitable for most paints but the paint must be thinned before spraying.

Airless spray.

Again suitable for most paints but while some paints must be applied by airless spray, others are unsuited to this method. While it is probably the fastest method of all, it is also the most complicated as different paints require different sized nozzles. Paint is delivered to the gun by means of an air or electrically operated fluid pump. Once again the paint is normally thinned before spraying.
NEVER ATTEMPT TO OPERATE AN AIRLESS SPRAY GUN UNLESS YOU HAVE BEEN THOROUGHLY AND PROPERLY INSTRUCTED IN HOW TO USE THE GUN.
WARNING: The actual fluid pressure at the gun may be as high as 3,000 p.s.i. For this reason caution must be used in handling an airless spray gun and it MUST NOT BE AIMED AT ANY PART OF THE BODY under any circumstances.
CAUTION: Because of a static electricity potential generated by the pressures necessary for airless spraying, it is possible that sparking may occur between gun and the object being sprayed. This can result in an explosion and/or fire. BE SURE THAT THE OBJECT BEING SPRAYED AND THE AIRLESS EQUIPMENT ARE GROUNDED. This can be done by attaching a static wire to water piping, electrical conduit, or any structural member known to be grounded. If the hose does not contain a static electricity conductor, a static wire must be attached from the spray gun to a suitable earth connection.

When spraying with either a conventional spray or airless spray, keep the gun perpendicular (at right angles to) to the surface being sprayed at all times. Do not arc the gun. Arcing leaves an uneven coat of paint. All spray guns must be properly and thoroughly cleaned immediately after use according to the instructions.
IMPORTANT: Spray guns are not intended for use with highly corrosive, highly rust inducing or highly abrasive materials.

Some paints contain skin irritants others contain toxic vapours, some may even contain both. Any instructions concerning the wearing of goggles, gloves

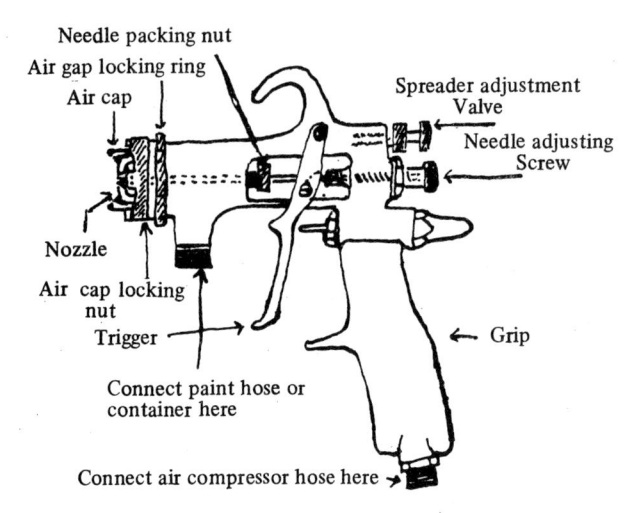

Needle packing nut
Air gap locking ring
Air cap
Spreader adjustment Valve
Needle adjusting Screw
Nozzle
Air cap locking nut
Trigger
Connect paint hose or container here
Connect air compressor hose here
Grip

A Typical Modern Paint Spray Gun

SPRAYING

THIS

NOT THIS

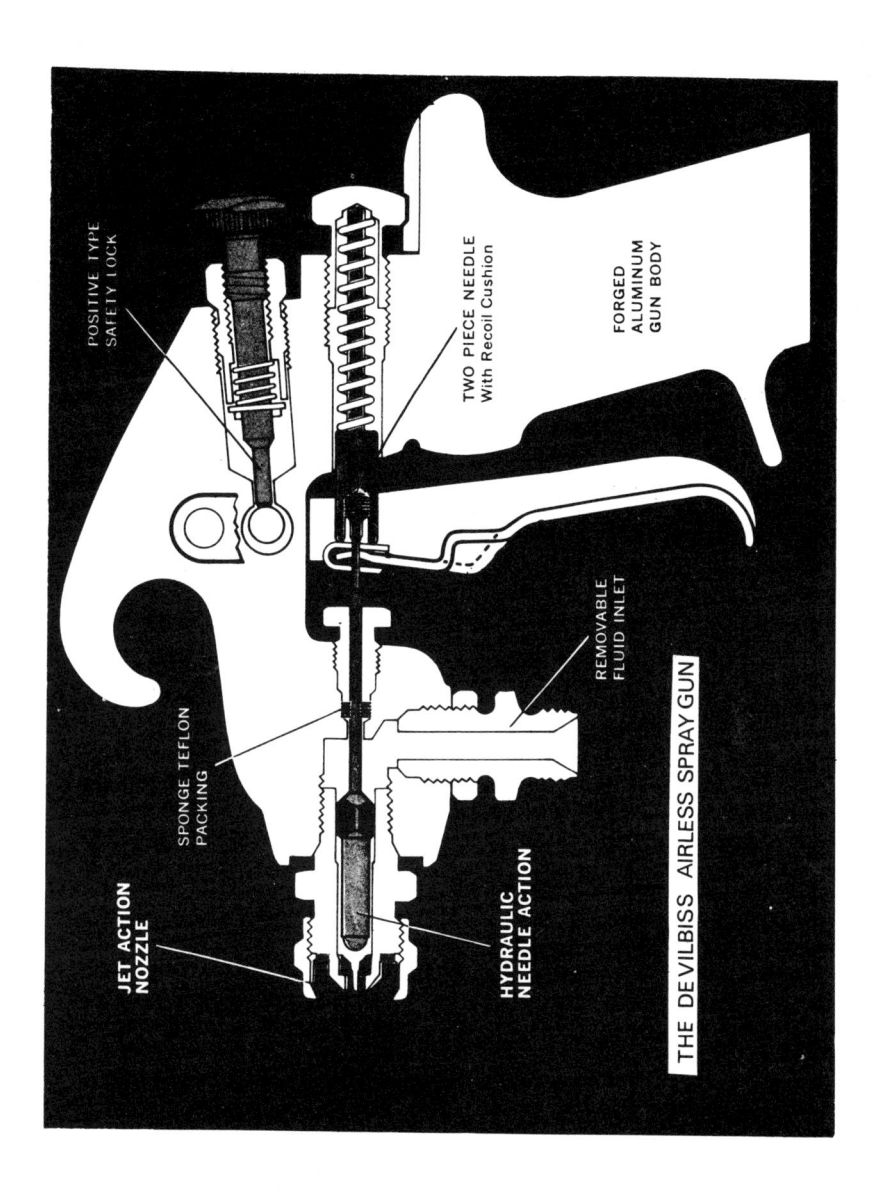

- POSITIVE TYPE SAFETY LOCK
- TWO PIECE NEEDLE With Recoil Cushion
- FORGED ALUMINUM GUN BODY
- REMOVABLE FLUID INLET
- JET ACTION NOZZLE
- SPONGE TEFLON PACKING
- HYDRAULIC NEEDLE ACTION

THE DEVILBISS AIRLESS SPRAY GUN

and protective clothing should be strictly adhered to. When the paint contains toxic vapours breathing apparatus must also be worn, most especially when painting in poorly ventilated spaces.

Woodwork too requires constant attention. Decks are best cleaned by removing the surface dead wood with holy stones. However, this is ardous work and today a corrosive is often used for this purpose. When using a proprietary cleaner, suitable footwear should be worn and the hands kept away from the cleaner. If the flesh is splashed it should be well washed with clean fresh water immediately.

Bare wood which is exposed to the weather gives up its natural oils. It has always been considered good practice to coat wood decks and other exposed bare wood occasionally with a thin coating of raw linseed oil, to which a little red lead is sometimes added. Teak rails used to be kept spotless by the frequent application of sand, canvas and elbow grease, while other outside woodwork is either painted or varnished. Woodwork within the accommodation is normally painted with a fire retarding paint, always excepting furniture which used to be stained and polished. Today much of the interior woodwork is covered with a suitable plastic such as formica. Wood hatchboards may be treated with a proprietary wood preservative, other woodwork contained in the holds such as cargo battens, limber boards, ceilings and shifting boards usually remain untreated.

Blistering of paintwork whether on steel or wood is usually caused by moisture beneath the paint or sometimes on wood by the escape of resin from the wood. New wood which it is intended to paint should always be knotted and primed first. Both steel and wood surfaces should be quite dry.

Tools.

Always use the right tool for the job, keep all tools clean and stow them back in their proper place after use. When sending any tools down a hold, tank or other compartment, put them in a canvas sack so that they cannot drop out. When it is necessary to use tools in an atmosphere that could possibly be combustible, the tools are to be of bronze or a similar metal in order to avoid any possibility of sparking.

Before using a spanner, find out both the size and type you need, also the thread on the bolt. Thread may be Whitworth, British Standard (B.S.), American (across face, A.F.) in which case the size will be in inches and they are not interchangeable or in Metric (continental) in which case the size will be in millimetres and once again it will not be interchangeable with the others. Types include open-ended, ring, box and socket. Adjustable and shifting spanners are supplied for use when the correct spanner is not available but these spanners should not be used if it can be avoided. Use a proper wheel spanner on valve wheels.

The same thing applies to screwdrivers of which there are three types. The normal flat chisel ended screwdriver for British screws, the Phillips screw driver which fits into a star in the head of the screw and was originally American and an Allen key which fits into a six sided hole usually in a bolt head.

Pipe wrenches, pliers and grips of various types should never be used on nuts or bolt heads, they round off the corners and make it impossible for a spanner to grip.

Always use a podger (small crow bar) when tightening bottle screws.

To avoid corrosion, always oil or grease bolt threads before replacing nuts and put a little vaseline on the threads of wood screws.

Whenever electric or air powered tools have been used, they should be dismantled according to the instructions, cleaned and where necessary oiled, before being stowed away, examine all electric wiring for any damage that may have occured and if damaged it must be renewed before being put into use again. When using air powered tools, the air jet must not be allowed to strike any part of the body.

Oil lamps. Every ship is required to carry spare oil N.U.C. and Anchor Lanterns for use in an emergency. To trim an oil lamp;—Fill the container with paraffin, using a funnel, then replace wick and burner. Turn up the wick and using the thumb and finger, remove the crust from the top of the wick. With a pair of lamp-trimming scissors cut off any loose threads of wick and round off the corners. Clean and replace the chimneys. The lamp is now ready for use. Colza oil lamps are used in binnacles and most lifeboats. **Do not** put paraffin in a colza oil lamp.

Rounds. Every four hours before going off watch at night, one member of the watch should go round the decks and public rooms, to ensure that all is well. He should take a torch, hammer and when there is a deck cargo, a podger. All wood hatch wedges and deck cargo bottle screws should be examined for tightness and made tight wherever slackness has occured. Rope lashings on stores and deck cargo are to be examined and any slackness taken up. The stern light is to be sighted to ensure that it is alight. Rod and chain steering gear when fitted is to be closely examined and where a patent log is streamed, it is to be read. He will also keep his senses alert for any smell of smoke or burning, especially when passing ventilators. He will then report to the Officer of the Watch before handing over to his relief and going below (off watch).

Hygiene. At sea all rubbish and swill to be thrown over on the lee side. In port it is to be placed in the receptacle provided by the Port Authorities. It is an offence to allow anything to go overside in most ports, particularly if they are dredged or the water is stagnant. In most ports, unless the ship is equipped with a septic tank capable of retaining all sewage, the ship's toilets must not be used and the crew are required to use the shore facilities provided. Portable or permanent overside discharge covers are to be placed over discharge pipes to ensure that any condenser circulation water, bilge water, ballast water, (and if the ship's toilets are in use, sewage water) or any other water, is not allowed to flow over the quay or boats or barges or other craft tied alongside. The discharge of oil is strictly prohibited in all ports.

When washing down with the hose, it is necessary in some ships to have at least one cock on the deck service line open all the time the pump is on. Therefore under these conditions it is necessary to open a cock on the deck service line before shutting off the hose. Always have a second man to light the hose to the man attending the nozzle. In port, if washing down, it may be necessary to block certain scuppers before putting the hose on, so that the waste water does not go either onto the quay or into a craft alongside.

Never leave oil or grease laying about either on the decks, work-benches, store-rooms or anywhere else, wipe it up. Do not let wet or oily rags accumulate anywhere, dispose of them.

Wire ropes. To reeve a new runner when the derrick is topped. Take the old runner off the winch and out of the heel block. Marry the bare end to the bare end of the new runner and pull through the head block. Take the end through the heel block and clamp to the winch barrel. Hold the runner over the far side of the barrel from the clamp and let the turns cross the end of the runner as they are wound on the barrel. A swivel in the head block will allow it to turn.

To oil a mooring wire, take all the wire off the reel. Lay a piece of old canvas on the deck close to the reel. Apply fish oil to the wire with a wad of old rag, reeling the wire up as you go. Wear gloves.

Cleaning brass. To clean brass that is coated with verdigris. Bind a bundle of rope yarn tightly and cut it close to the binding to make a brush. Dip the rope yarn in salt and lemon juice and rub the brass briskly. When clean, wash with fresh water and polish in the normal manner. If the brass is in an exposed position, coat with colza oil after cleaning.

Canvas. When sowing or repairing canvas, pull the seaming twine over a piece of bees wax both to prevent kinking and as a preservative. To paint new canvas, stretch it taut in position, wet it and paint while wet.

Cement. Keep dry. To use, mix 1 part cement with 4 parts fresh water sand, use fresh water.

Petty pilferage. Before entering port remove all brass sounding pipe caps and plug the pipes with wood plugs. Replace caps when the ship sails. Keep all accommodation doors LOCKED while the ship is in port.

Soundings. All peaks, bilges, double bottom tanks, etc., are to be sounded twice daily whenever the ship is at sea. When using a wet sounding rod, rub chalk along the rod to obtain a sounding. (depth of water in the tank).

Oil Bunkers. When bunkering, all scuppers from which oil could spill must be blocked.

ACCIDENT PREVENTION.

NEVER sit on the bulwarks or rails.
NEVER go outside a lifeboat if there are no rails there, except for the purpose of carrying out your duty.
NEVER wear footwear with steel tips on the soles or heels.
NEVER leave a door loose at sea, close it or put it on the hook.
NEVER jump from ship to quay or quay to ship. Use the gangway.
NEVER carry anything in your hands when climbing a ladder. Pull the gear up afterwards. Do not put it in your pocket, where it can fall out.
NEVER carry anything in such a way that you cannot see where you are going.
NEVER work on the funnel or in the vicinity of the whistle unless both the bridge and engine room have been informed.
NEVER work in the vicinity of a radio aerial or lead-in unless the wireless room has been informed.
NEVER work overside in the vicinity of the propeller unless the engine room has been informed.
NEVER leave any rope laying around loose. Coil it down.
NEVER rub down old paint containing lead unless you are wearing a suitable breathing apparatus.
NEVER smoke while painting in a confined space, or until the paint is dry.
NEVER hoist or lower a drum by a handle—Use a strop.
ALWAYS stow casks and drums "Bung Up."
When painting with an airless spray, a loose sleeve of old air hose about 10 feet (3m) long should be slipped over the line adjacent to the gun and paint container.
When using a punt for painting overside, do not rig a stage to gain height. Make sure your painter is strong enough to hold the punt against any tide or current. Float a lifebuoy astern.

L.P.G. Vessel M.V. Wiltshire Messrs. Bibby Lines Ltd. Showing pipe arrangement on deck for loading and discharging liquid gas cargoes.

CHAPTER 13.
HAZARDOUS CARGOES.

Ammunition will normally be carried in a specially constructed magazine. Apart from smoking and handling restrictions when loading and discharging and in the vicinity of the magazine, throughout the voyage. The carriage of this type of cargo will not normally affect the working of the ship.

Corrosive and/or toxic fluids such as acids may possibly be carried in properly constructed small containers as deck cargo. In the event of one or more containers becoming damaged and possibly leaking, the damaged containers will have to be jettisoned (thrown overboard). Proper protective clothing (particularly eye and facial) and if necessary suitable breathing apparatus must be worn while the work of jettisoning the containers is carried out. A substance suitable for killing any spilled chemical and an antidote suitable for application to the person, should be carried on the ship. All such containers are to be stowed bung up and must be properly secured and as far as possible protected from damage.

Inflammable liquids in suitable containers may be carried below decks under stringent regulations which are detailed in "The Carriage of Dangerous Goods in Ships" otherwise known as "The Blue Book".

Oil, Liquid gas and Chemicals of various types are carried in bulk in properly constructed ships. The main hazard, is equally present not only all the time the cargo is on board, but all the time the ship is in ballast as well, and is to be found in the combustible and/or toxic gases and vapours given off by the cargo and when the ship is in ballast, the cargo residue. Even when the ships are in ballast and the tanks have been washed and gas-freed, the residue of cargo remaining in the sludge and scale on the interior of the empty tanks can give up a large and dangerous volume of either gas or vapour and cause the atmosphere in a tank, or pockets of atmosphere in a tank, to again become combustible and/or toxic, most especially when climatic temperature increases. Obviously then, in order to preserve both the ship and the lives aboard her, it is necessary to take continual precautions to try and ensure that (a) Personnel do not inhale toxic gas and (b) combustible gas is not ignited. No member of a tanker's crew should ever assume that the ship is entirely gas free.

To try and explain some of the measures that are taken to ensure safety of both crew and ship, it is probably best to follow a ship on a round voyage, commencing at the loading port.

The ship when she arrives at her loading port, will be in ballast. Some of the tanks will be empty and ready to receive cargo, some may possibly contain sea water ballast. In the empty tanks, cargo residue may have given off a certain amount of gas or vapour. It may or may not be sufficient to make the atmosphere, or part of the atmosphere, in the tank explosive or toxic. Petroleum vapour and many others are heavier than air, so the vapour has probably concentrated on the bottom of the tanks.

Before a vapour can ignite, it must have a required percentage of oxygen mixed with it (Too much or too little oxygen will make the mixture either too rich or too weak and the mixture will remain non-combustible). The mixture will not ignite below a certain temperature but as the flash point of most vapours is very low, this is hardly likely to influence the position. Should the temperature be sufficiently high, the vapour will ignite spontaneously but again the temperature required for this is too high for it to be a likely possibility.

The Cowal swivel hatch cover for tankers can be fully opened in 8 seconds or locked in a partially open position. Fully oil/water tight.

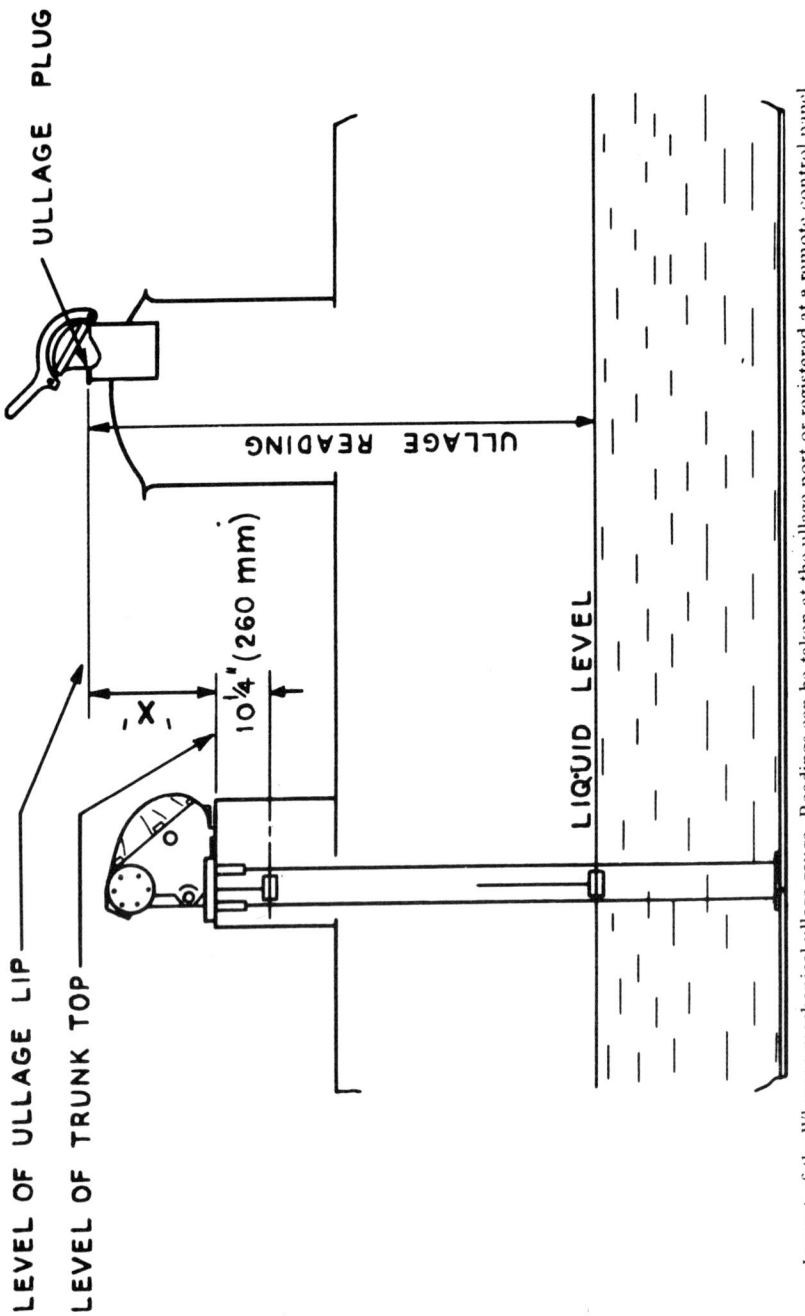

Layout of the Whessoe mechanical ullage gauge. Readings can be taken at the ullage port or registered at a remote control panel.

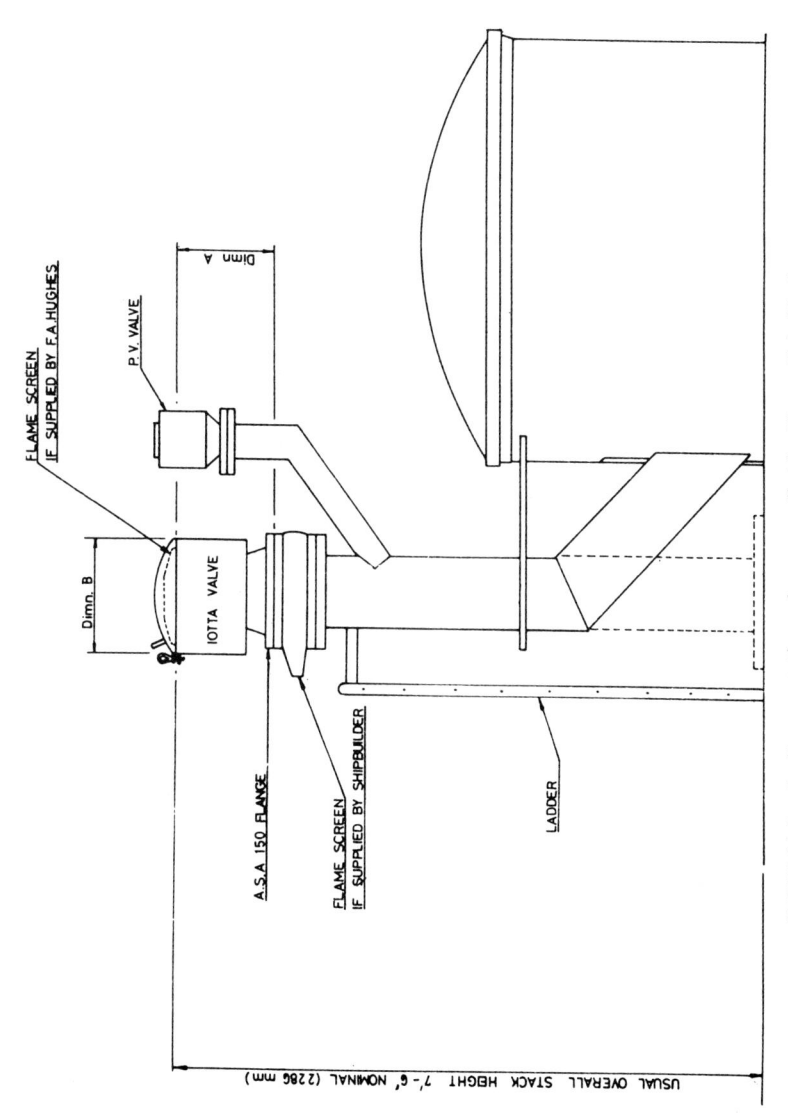

IOTTA high velocity gas venting valve as supplied by Messrs F.A. Hughes.

FIGURE 1
INLET PRESSURE LOWER THAN SETTING

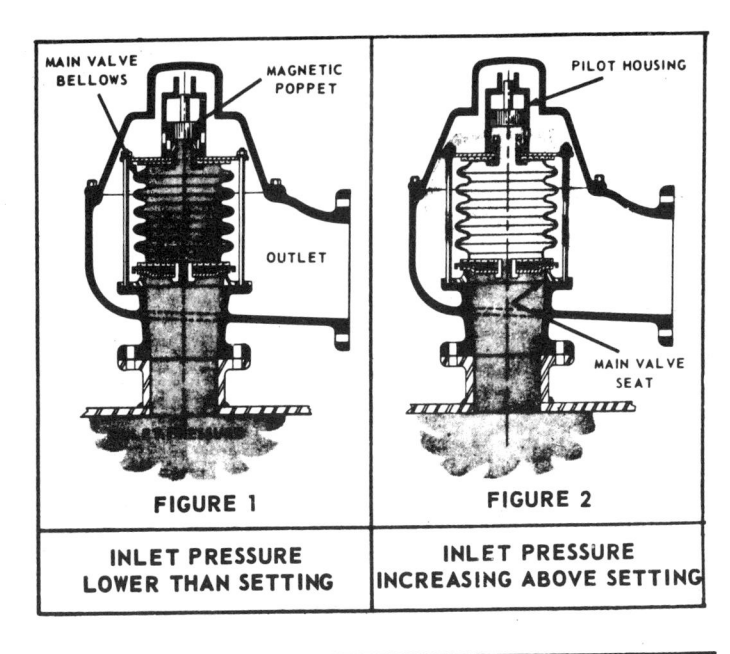

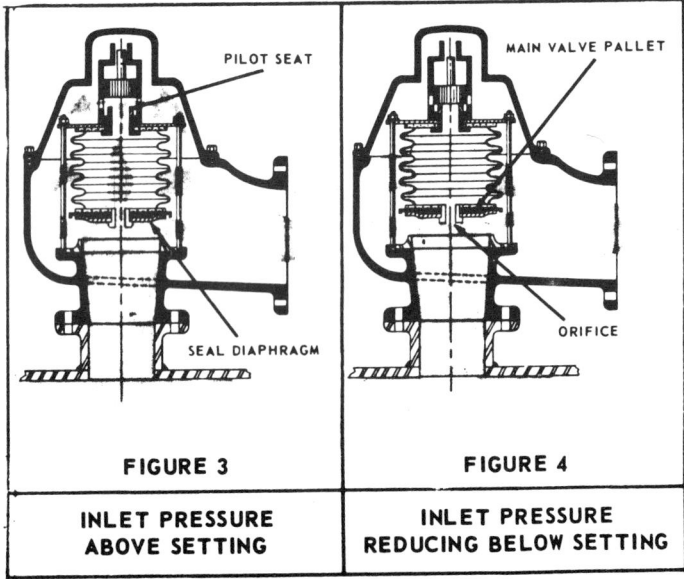

FIGURE 2
INLET PRESSURE INCREASING ABOVE SETTING

FIGURE 3
INLET PRESSURE ABOVE SETTING

FIGURE 4
INLET PRESSURE REDUCING BELOW SETTING

Whessoe Figure 4610 Pilot Operated Safety Relief Valve.

Fig. 2

Gas exit rate up to 300ft/sec. (92 metres/sec.)

Water tight hatch.
— Opened when loading

Vent aperture opens at 0·25lbs/sq.in. (0·018kg/sq.cm)

MARTIN HI-JET SAFETY VENT Constructed from non-corrosive material with anti-static properties.

EXPANSION TRUNK

MAIN DECK

Pressure and vacuum breather

Water tight door Opened when discharging

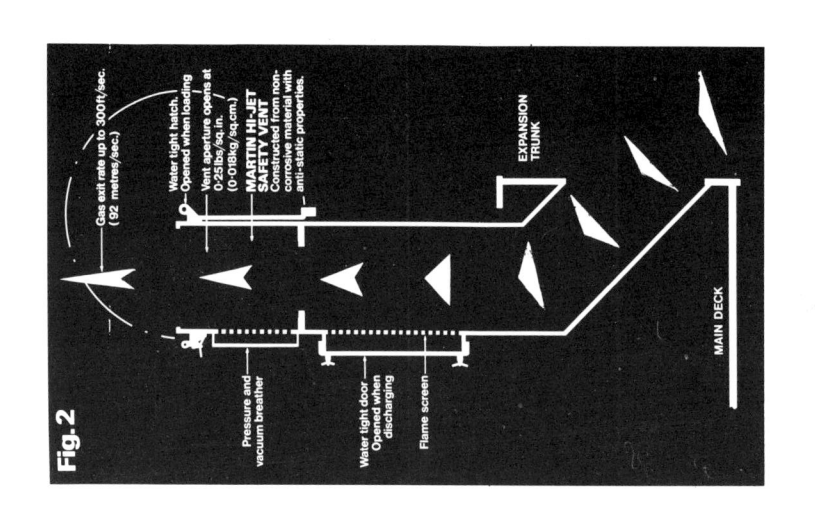

Flame screen

Fig. 1

MARTIN HI-JET SAFETY VENT

Therefore if the vapour is to catch fire (and if it does, this may well result in an explosion) it must be ignited and precautions have to be taken to prevent this happening.

Apart from the obvious causes of ignition such as smoking, there is another cause which is much harder to control i.e., sparks. Sparks may be produced in the funnel, or by dropping a metal tool on the deck, or by the flint in a lighter, or by the switching on of a torch or electrical apparatus. They may also be caused by the earthing of static electricity which has built up anywhere, in a metal ullage tape, the human body, droplets of water mist or steam or elsewhere. Static sparking is probably the most difficult ignition source to control.

The concentration of electricity in the hull of the ship may well be different to the concentration ashore. To avoid sparking when hoses are put aboard, and a connection made between ship and shore and to make the charges even and obviate any chance of sparking between the ship's hull and the shore. A good stout earthing wire is sometimes connected between ship and shore, before loading (or discharging) of cargo commences.

The cargo loading hoses, which also normally contain an earthing wire, are then connected up so that cargo can be loaded. The connecting of the hoses is usually done by the crew. The hose connections may have raised flanges and bolt holes, or a speed coupling may be used. The phosphor bronze (or other material) tools supplied are to be used to make the couplings and care must be taken to ensure that the earthing wire remains continuous and that static electricity is not given an opportunity to jump from hose to ship or ship to hose. Also ensure that the jointing and all flange faces are spotlessly clean. Any particular instructions that have been issued regarding the sequence of coupling the hoses is to be strictly adhered to. Instructions issued regarding the wearing of protective clothing, particularly eye and facial protection, are to be meticulously observed, no matter how irksome they may be or how small the risk is considered to be.

Depending upon the type of cargo that is to be loaded, it may be loaded by either (a) Open system or (b) closed system. As the cargo is pumped into a tank, the atmosphere in the tank must be discharged, in order to make room for the cargo.

When the cargo is considered to be safe: —That is to say that the gas or vapour which will escape from the tank as it fills, is not considered to be either explosive, combustible or toxic. The open system of loading may be used, this allows ullage ports and vents to be left open so that the atmosphere in the tank escapes at normal atmospheric pressure at deck level as the tank fills.

When however, the escaping gas or vapour is considered to be either combustible or toxic, the closed system will be operated. All hatch covers, ullage ports and other deck openings are to be kept tightly closed and the atmosphere from the tank will only be allowed to escape through pressure vents, which may be up the mast or may be at deck level. As cargo is pumped into the tank, the atmospheric pressure in the tank increases. When this pressure reaches a pre-determined level, the pressure vents will open and the gas or vapour will be forced into the open at speed, escaping rapidly from the vicinity of the ship and dispersing harmlessly in the open air. When the pressure is reduced to a pre-determined level, the valve closes until the pressure has built up again. Under these conditions, no member of the crew should ever come in contact with either the cargo or the atmosphere escaping from the tanks.

At one time, vents were simply led up the mast and the gas or vapour was allowed to escape at atmospheric pressure and was expected to disperse in the air

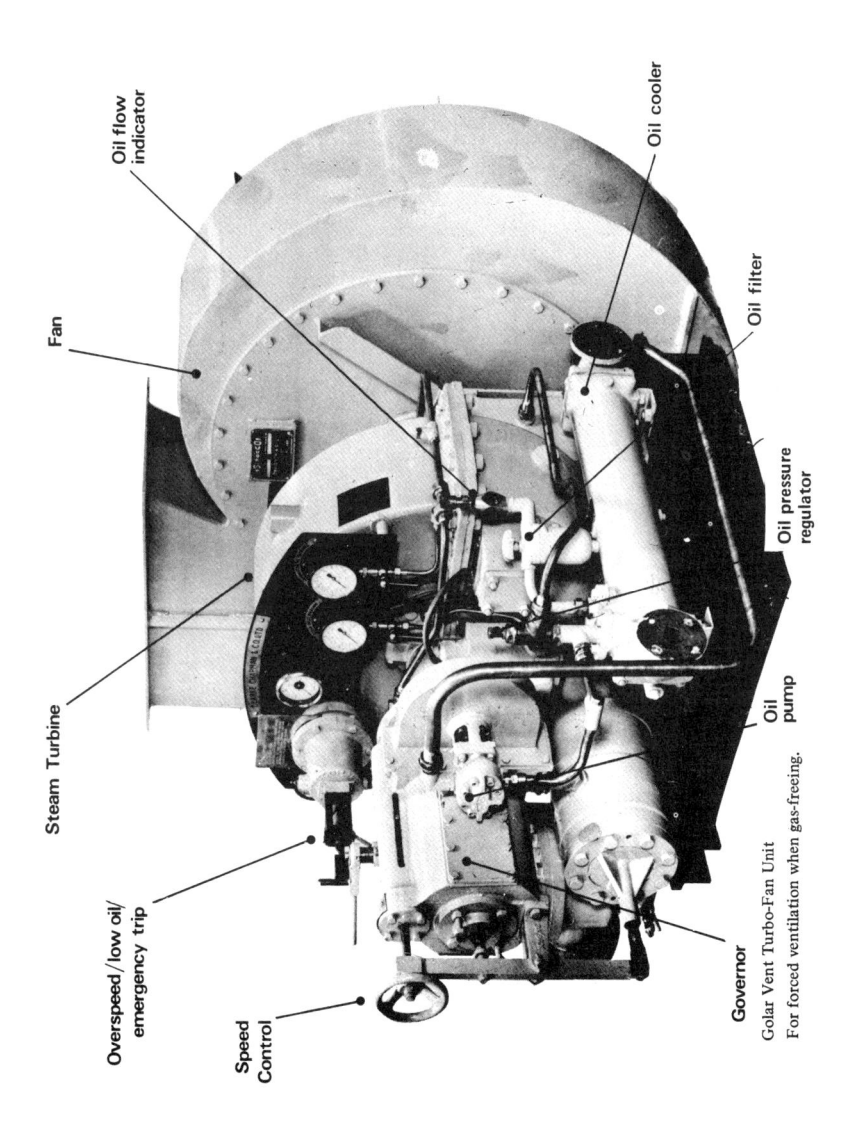

- Oil flow indicator
- Oil cooler
- Oil filter
- Fan
- Oil pressure regulator
- Oil pump
- Steam Turbine
- Overspeed/low oil/ emergency trip
- Speed Control
- Governor
- Golar Vent Turbo-Fan Unit
  For forced ventilation when gas-freeing.

F. A. HUGHES INERT GAS SYSTEM

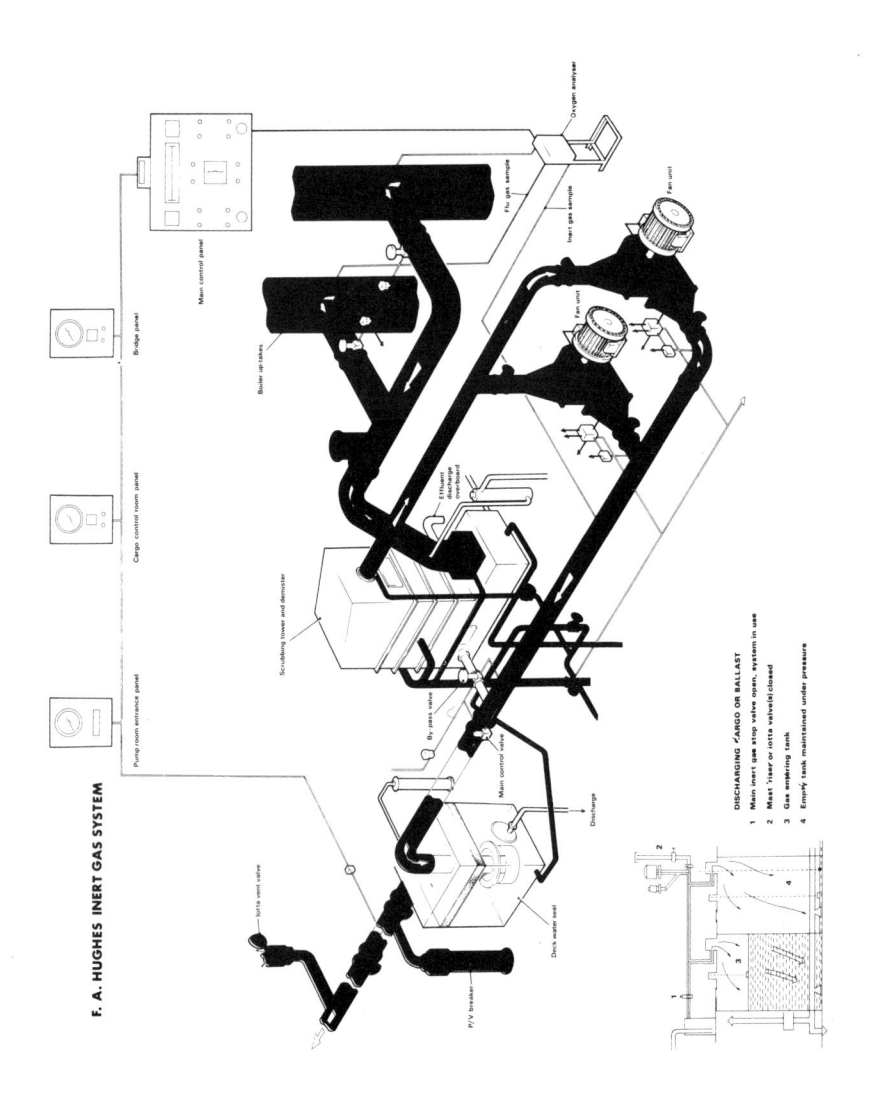

DISCHARGING CARGO OR BALLAST
1. Main inert gas stop valve open, system in use
2. Mast 'riser or iotta valve(s) closed
3. Gas entering tank
4. Empty tank maintained under pressure

GOLAR EJECTORS

APPLICATION.

TANKERS: Stripping cargo, clean ballast or machine washing residue. Chain locker, fore peak, pump room and engine room bilge stripping.

BULK CARRIERS AND DRY CARGO VESSELS: Stripping ballast and cargo hold washing residue. Deep tank stripping. Engine room bilge, fore peak and chain locker stripping.

TYPICAL ARRANGEMENTS.

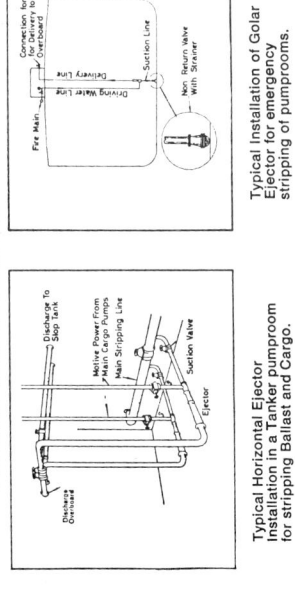

Typical Horizontal Ejector Installation in a Tanker pumproom for stripping Ballast and Cargo.

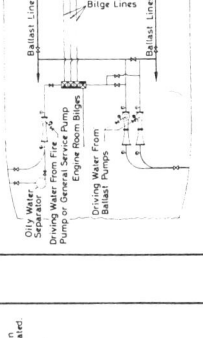

Typical Installation of Golar Ejector for emergency stripping of pumprooms.

Arrangement of Ejector for stripping chain locker.

Golar Ejector rapid ballasting and deballasting arrangement (Tanker ballast only compartment).

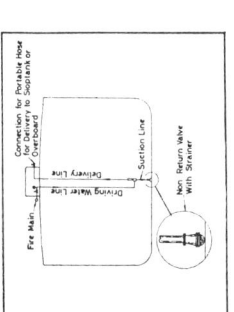

The diagram shows how installing Golar stripping Ejectors eliminates the need for conventional bilge pumps on Bulk Carriers.

above the ship. However, toxic and combustible gases are more often than not, heavier than air. When the weather is calm, gas or vapour that is not expelled under pressure tends to fall, with the result that it can collect at deck level and large pockets of gas are prone to collect in the lee of the deck superstructure, with what could be disastrous results. Nevertheless, whether or not the gas or vapour is escaping at atmospheric pressure or under pressure, all doors, ports and other openings facing the main deck should be kept closed in order to prevent any pocket of gas or vapour, which may have collected under the lee of the accommodation, from entering and spreading through the accommodation.

As the cargo in the tank increases in volume during loading, it will continue to give off more gas or vapour, which in turn will increase the density of the atmosphere in the tank. With many cargoes the atmosphere in the tank (as the cargo is being loaded) will become first of all explosive and then too rich to ignite, until it has been again diluted with the oxygen contained in the fresh air.

As each tank fills, the cargo coming aboard by hose, will have to be diverted to another tank. The various valves on the main deck will be coloured to indicate their various functions. Because this colour code can and does vary from company to company. No member of the crew should ever attempt to open, close or in any way interfere with any valve, without orders being given to him by a responsible officer and he should be quite certain which valves he has to open, which valves he has to shut and in what order the opening and shutting of the various valves is to be carried out. Opening or closing the wrong valve can cause a disaster.

When the cargo has been loaded it will be necessary to disconnect the hoses. Disconnect in the correct sequence in order to ensure that static sparking does not occur. Care must be taken to ensure that the hoses are empty. The lower bolts should always be released first and protective clothing must be worn, particularly eye and facial protection.

Ullages (measuring the amount of empty space above a cargo in a tank) will usually be taken by mechanical means. Some may however need to be taken by hand, using an ullage tape which is let into the tank by means of an ullage port (a special opening in the deck supplied for this purpose). When a metal ullage tape is used, attention must be paid to the instructions for use and to detail, in order to avoid any risk from sparking caused by static electricity either in the tape or in the body of the crew member.

Once the ship is loaded and away, increases in climatic temperature can cause an excess of gas or vapour to concentrate under pressure in one or more of the tanks. This pressure has to be released either by natural or pressure venting.

On arrival at the port of discharge, the whole process of loading goes into reverse and air enters the tanks as the cargo is discharged, either naturally or by means of vacuum valves incorporated in the pressure valves. The over rich gas filled atmosphere in the tanks, will quickly become thinned and possibly explosive or combustible. To overcome this hazard, many ships now produce or manufacture an inert gas, which is fed into the tanks in place of atmosphere, as required. The consequent lack of oxygen in the tank, when it is fed with an inert gas, ensures that the atmosphere in the tank cannot become combustible or explosive. It will of course be highly toxic.

The main suction lines employed to discharge the cargo, have a large bell mouth suction and are therefore unable to cope with the last of the cargo remaining in the very bottom of each tank. These tank dregs are removed by a smaller stripping pump, which has its own separate suction line.

When the cargo has been discharged and the ship proceeds to sea, any tanks which have not been filled with water ballast will (unless an inert gas has been used to replace the cargo) contain a high percentage of gas or vapour which may well be combustible. Tanks that have been filled with water ballast, will of course have had to have the gas or vapour vented as the tank was filled with ballast.

Unless the tanks merely contain an inert gas and the next cargo is to be of the same grade as the last one and there is no scale in the bottom of the tank. The first job will be to gas-free and wash the tanks. The Dover covers are taken off the butterworth holes in the deck for gas-freeing by Extractors and inserting tank washing machines, coupled to hoses, which are then lowered into the tanks (one or two tanks being treated at a time) by means of a fibre rope secured to the tank washing machine for this purpose. To avoid sparking, the hoses which contain an earthing wire, must be connected in the correct sequence and manner. Some ships have fixed washing machines in the tanks, however, these are not always used, because there is a possibility that fixed washing machines may be more inclined to induce the sparking of static electricity, than the portable ones. On small ships the hosing may be done by hand.

Because static electricity would seem to build up more easily in hot water and steam, the first wash is normally with cold sea water. The tank washing machine is operated at different levels in the tank (being raised or lowered on its own fibre rope) to ensure that the tank is well washed. Stripping pumps or portable ejectors are used to remove the tank washings, which may all be pumped into and retained in one tank, or in the open sea pumped overboard. However, in view of the pollution pumping oil overboard helps to cause, many ships are now first removing all the oil and sludge from tank washings, before pumping the water overboard. According to the nature of the previous cargo and the next, tanks may be washed with cold salt or fresh water, hot salt or fresh water, or steam, as required. After washing and stripping (pumping out) the tanks must be gas-freed again by ventilation. Small ships may use windsails to do this but the method is too slow for large tankers which will use either fixed or portable fans and or extractors to give forced ventilation.

When the tanks are gas-freed, it may be necessary to enter one or all of them, for the purpose of removing sludge and loose scale. This is necessary to ensure that the suctions do not get clogged when next discharging. No person is to enter any tank before the atmosphere has been tested and pronounced gas-free. Since gas and vapour can accumulate in pockets, a number of atmospheric tests should be taken from different parts and at different levels of the tank. It is essential that all men entering a tank wear lifelines and, if necessary (depending on the nature of the previous cargo) protective clothing as well. An attendant (who should have suitable breathing apparatus available) must be stationed at the entrance, in a position from which he can see all the men in the tank, for the purpose of raising the alarm if he should have any cause to think that any of the men in the tank are in any danger of being overcome by fumes.

A rise in the temperature of the atmosphere in the tank or the removal of sludge or scale, may well release additional gas or vapour and can very quickly turn a safe atmosphere into a toxic one, which could easily result in the men in the tank being overcome without they themselves appreciating that it was happening.

Victor Pyrate Dover cover

Gas freeing a tank with Marland Vacuum extractions.

Used in pairs one extracting and one venting the vents are spark proof and non-static. light and easy to handle, with no moving parts. If damaged can be repaired with a fibre-glass repair kit.

The Mines Safety Appliance Coy's Lamb Air-Mover. Portable blower or exhaust unit for removing hazardous concentrations from confined areas or for cooling atmospheres in hot operations.

The Victor Pyrate standard hose saddle for use with portable tank washing machines

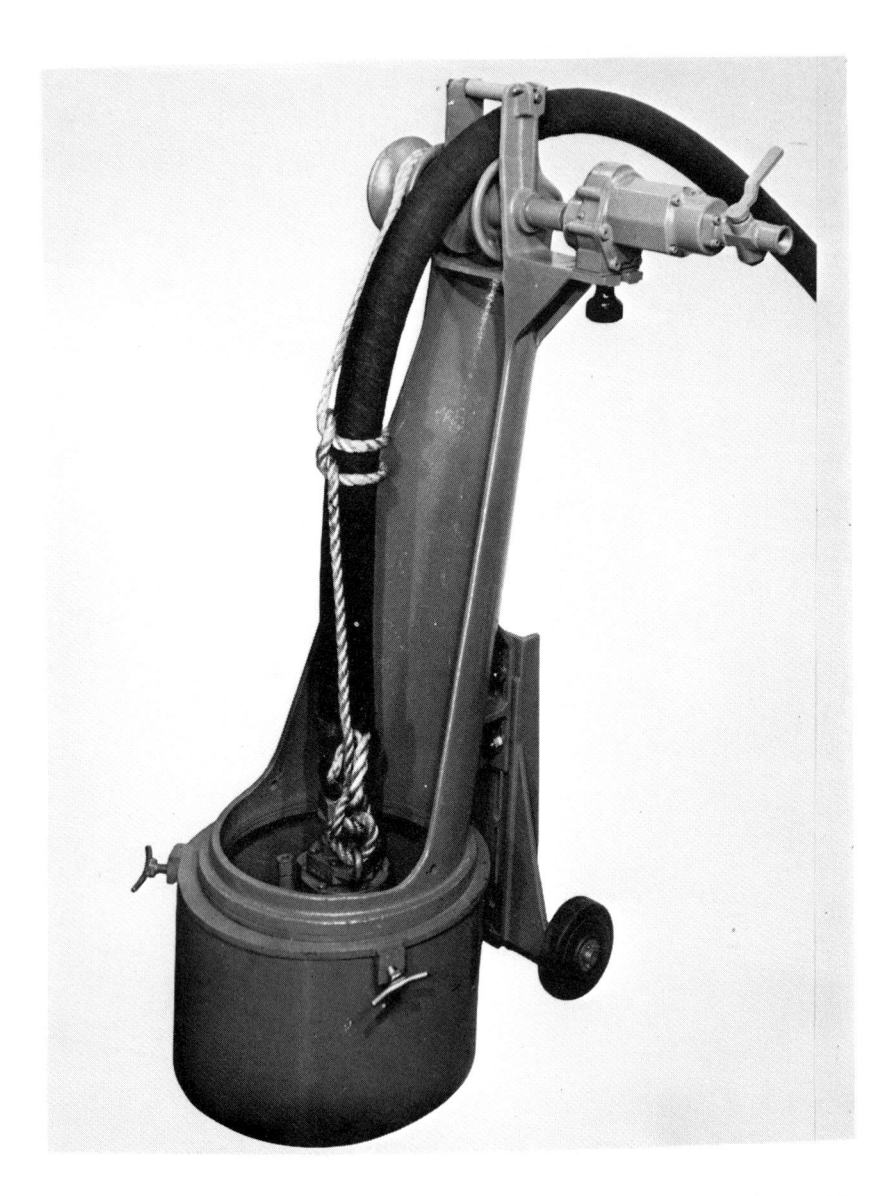

The Victor Pyrate pneumatic hoist for tank washing machines in position with a machine ready for lowering into a tank.

A conventional Victor Pyrate portable tank washing machine

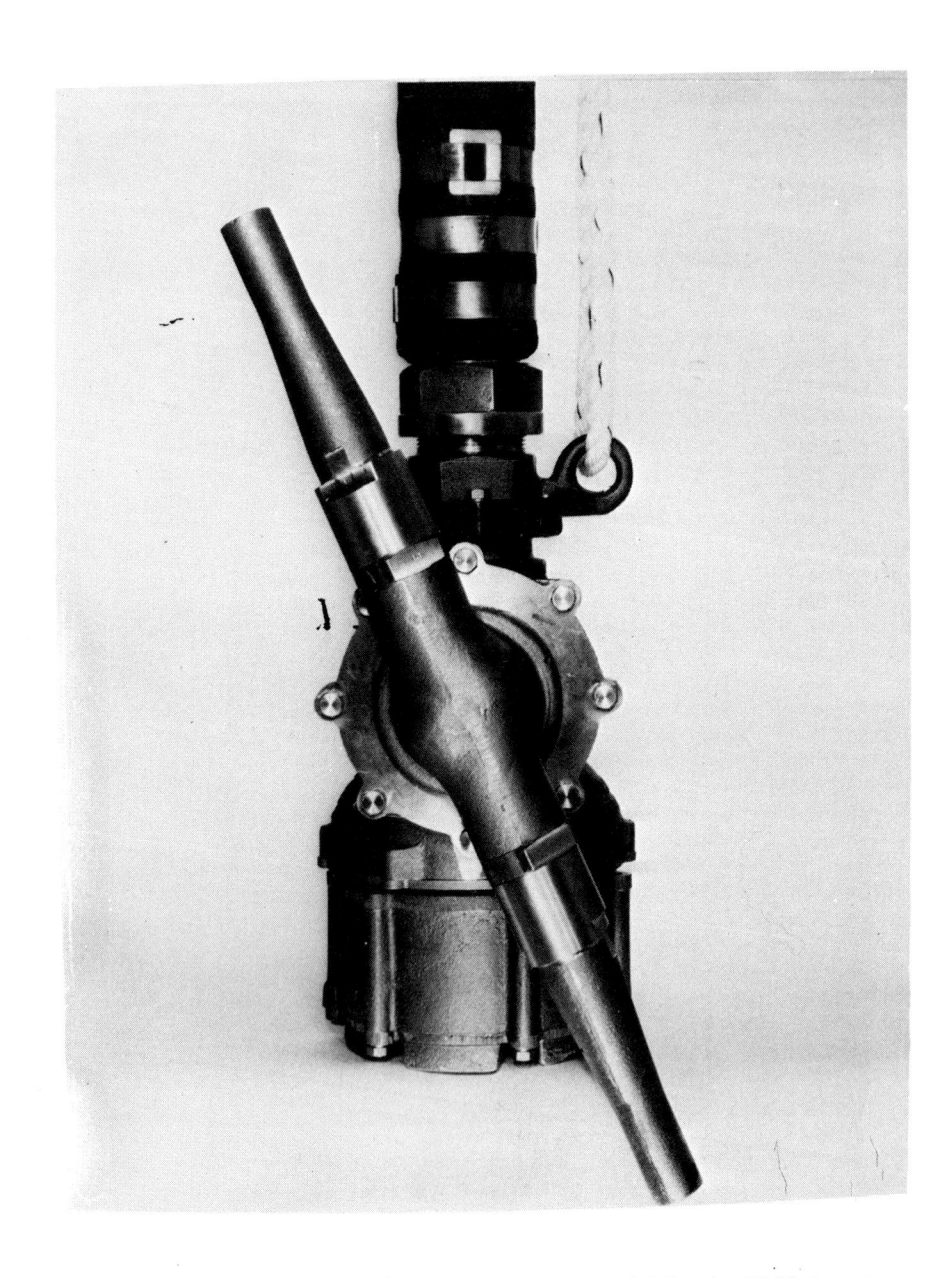

A Victor Pyrate major protable tank washing machine intended for the VLCC tonnage having two larger nozzles.

The Victor Pyrate portable tank bottom washer

V.P. Matic installed tank washing units

Wood or plastic shovels and spades should be used for the removal of sludge and scale which may be sent up in a plastic bucket. Any tools that have to be used should be of bronze or some other material that cannot cause sparking. All tools are to be lowered into and hove out of tanks in a canvas bag or holdall. They are not to be stuck in pockets or belts, or be thrown.

Similarly, no pump room or other confined space is ever to be entered until (a) the atmosphere has been properly tested and (b) the permission of a responsible officer has been obtained.

Care and maintenance of tank washing hoses.
1. Lay straight on racks if possible and store in cool conditions out of direct sunlight. Do not overstow.
2. Flush through with clean water if used with chemicals.
3. Wipe clean of contamination after use.
4. When in use avoid sharp bends in hose and use saddle or hoist over tank opening.
5. Avoid contact with hot pipes, abrasive surfaces and sharp edges, etc., liable to damage outer cover.
6. Clean and grease coupling threads from time to time.
7. If built-in earthing circuit is broken this can be remedied by the use of a separate wire through the the bore of the hose. A flexible stainless steel wire should be used which must be properly attached to the couplings at each end of the length of hose.

SUMMARY

Do not smoke except in a permitted area and not even in a permitted area, if there is any danger of gas or vapour being present.
Do remember that you cannot always detect toxic fumes by smell.
Do not use any metal tools that can cause sparking.
Spark guards and flame arresters that have been removed for any purpose, are to be replaced as soon as the necessity for their removal has ceased.
Do not enter any tank or other confined compartment without the permission of a responsible officer, or without a lifeline and attendant.
Do not open or close any cocks or valves without first having received instructions to do so from a responsible officer.
Do not carry any matches, lighters or torches except those issued by the ship.
Do not go to the rescue of a man overcome by fumes, unless you are wearing suitable breathing apparatus and have an attendant standing by.
Do not attempt to go to any place or do any work, unless you are wearing the correct protective clothing, whenever protective clothing is considered necessary by the regulations, by an officer or by yourself.
Do not make or break any hose connection containing an earthing wire, except by observing the sequence laid down in the instructions and/or regulations meticulously.
Do not walk under flexible cargo hoses.
Know what action to take in the event of fire.
Know what to do if you are splashed by a corrosive or toxic liquid be it cargo or otherwise.
Know where the deck safety shower, eye wash spray and bottles, first aid kits, breathing apparatus and resuscitators are located.

Know that the proper tools must always be used on every job.
Help keep all safety equipment in first class condition.
Never leave equipment lying around, especially equipment that could cause sparks.
Fire hoses (with foam branches when applicable) should always be run out and ready before loading or discharging a highly inflammable cargo.

Merchant Shipping Notice No. M.576

CARRIAGE OF DANGEROUS CHEMICALS IN BULK

Danger from asphyxiation and/or affects of toxic or other harmful vapours on entry into tanks and other enclosed spaces

1. **General precautions**
   (a) When dangerous chemicals are being or have been carried in bulk, personnel should not enter cargo tanks, void spaces around such tanks, cargo handling spaces or other such enclosed spaces unless authorised by a responsible officer.
   (b) Even if there is no cause to suspect that such a place is not free of toxic vapours in harmful concentrations or deficient in oxygen, that officer should first check that the space is safe for entry by the use of gas indicators as described in paragraph 3 below and then ensure that :—
     (i) the space is effectively and continually ventilated
     (ii) a competent person is stationed at the entrance to the space to summon immediate help if that becomes necessary, and
     (iii) approved breathing apparatus (duly tested), lifelines and harnesses, are ready for immediate use.

2. **Emergency entry**
   Where entry into an enclosed space is necessary in an emergency and it is not possible to check the space with gas indicators as specified in paragraph 1(b) above, an officer should be responsible for continuous supervision of the operation and should ensure that:—
     (a) personnel entering the space are wearing breathing apparatus and lifelines, both checked to the satisfaction of the responsible officer
     (b) means of communication and a system of signals are agreed and understood by all those concerned, and
     (c) ventilation is provided if possible.
   (Note.—Anti-gas respirators of the canister type, where the atmosphere is drawn through an absorbent filter, should never be used in enclosed spaces.)

3. **Gas indicators**
     (a) Where the cargo is toxic, the enclosed space should be checked for vapour before entry and frequently re-checked during the operation. Special proprietary detectors specific for the particular cargo vapour and capable of detecting very low concentrations should be used: combustible gas indicators are not generally suitable for that purpose.
     (b) The atmosphere towards the bottom of the space should always be tested; also any suspected dead spots which might be less efficiently ventilated than others.

(c) Even when gas indicators are used which measure oxygen content, the need for thorough ventilation whenever personnel enter closed spaces is paramount.

4. Opening up equipment and fittings

Personnel should vacate any enclosed space into which liquid or vapour may be released as a result of a cargo pump, pipeline, heating coil valve, etc., being opened up. Re-entry should be allowed only on the authority of a responsible officer under the conditions outlined in paragraphs 1 or 2 above.

5. Sludge, scale, etc.

Personnel should not enter into any tank where sludge, scale, etc. might be present, and which could rise to harmful vapours when agitated, unless breathing apparatus is worn.

6. Warning notices

Suitable notices should be prominently displayed forbidding unauthorised entry into any enclosed space where there is a danger of asphyxiation and/or inhalation of toxic vapours.

7. Protective clothing

Where entry into an enclosed space may also present a skin contact hazard (i.e. liquid irritation or skin absorption), suitable clean protective clothing including goggles should be worn. Such clothing should also be made available for emergency use together with the other items mentioned under paragraph 1(b) above.

8. Explosion and/or fire hazard

This notice does not cover precautions to be taken to minimise the danger of explosion and/or fire on entry into an enclosed space; reference should be made to the Tanker Safety Code issued by the Chamber of Shipping of the U.K. relating to bulk carraige of petroleum and similar substances.

9. Merchant Shipping Notice No.M.467

Notice No. M467 deals similarly with precautions to be taken before entering tanks and other enclosed spaces and is superseded by this notice only insofar as bulk dangerous chemicals are concerned.

December 1969.

Merchant Shipping Notice No. M.1091

TRAINING OF CREWS OF PETROLEUM, LIQUID CHEMICAL AND LIQUEFIED GAS TANKERS

Notice to Shipowners, Ship Operators, Masters, Officers, Seamen and Marine Training Establishments

This Notice supersedes Notice M.771

Recommendations and advice on the training of masters, officers and seamen serving or about to serve in petroleum, liquid chemical and liquefied gas tankers are contained in Training Bulletin No. 1/81 issued by the Merchant Navy Training Board (MNTB). It is important that arrangements be made for all tanker personnel who require any of the training described in the Bulletin to undergo such training at the appropriate stage in their career or service in such tankers. Courses have been organised under the aegis of the MNTB, and directly by shipping companies under Departmentally approved training schemes.

All crew members joining a tanker for the first time
2. The Department recommends that all personnel should receive a basic introduction to the main hazards associated with the transport of petroleum, liquid chemical and liquefied gas by sea before joining ships carrying such cargoes or, exceptionally, on joining those ships. The Department, therefore endorses the MNTB recommendations that:
 (a) all crew members joining a tanker for the first time should receive basic safety induction training before engagement, or exceptionally as soon as possible thereafter, unless the individual has already received such training during a pre-season training course within the preceding five years;
 (b) crew members with less than two years previous service in tankers who have had a break from such service of more than five years should undergo this training on re-joining a tanker;
 (c) all crew members should attend the MNTB Stage I (two-day) fire-fighting course.

Officers and Ratings who are to have duties connected with the carriage of cargo with cargo equipment
3. The Merchant Shipping (Tankers—Officers and Ratings) Regulations 1984 provide that officers and ratings serving on petroleum, liquid chemical or liquefied gas tankers may only undertake specific duties and responsibilities in connection with the handling of cargo and cargo equipment if:
 (a) an officer's certificate of competency or service is endorsed with an appropriate dangerous cargo endorsement (see also paragraphs 8 and 9 below); or
 (b) other officers and ratings are qualified in accordance with those requirements of the Regulations described in paragraphs 4, 5 and 6 below.
4. These officers and ratings must either:
 (a) have served in the deck or engine department, as appropriate, in the

type of tanker concerned for at least six months during the five years before 28 April 1984; or

(b) have satisfactorily completed the MNTB Stage I (Two-day)* fire-fighting course; and

either

   (i) subject to the proviso in paragraph 5 below, not less than two months supervised shipboard service in the type of tanker concerned in order to acquire adequate knowledge of operational practices before assuming responsibility for their intended duties; or

   (ii) an MNTB tanker familiarisation course relating to the type of tanker concerned, together with not less than 14 days supervised shipboard service in a tanker of that type; or

   (iii) periods of familiarisation and instruction during not less than four loading or discharging operations, including at least one of each operation, on board tankers of the type concerned, together with a period of 14 days supervised shipboard service in a tanker of that type.

Evidence of attendance at fire-fighting and tanker familiarisation courses approved by the Government of a country outside the UK may also be acceptable, if regarded by the Department as of equivalent standard.

5. The period of two months supervised shipboard service specified in paragraph 4(b)(i) above is reduced to one month for those officers and ratings who have satisfied the requirements of either paragraph 4(a) or 4(b)(i) in another type of tanker.

6. During periods of supervised shipboard service referred to in paragraph 4 above not more than one deck officer and one engineer officer forming part of the regular crew complement should be engaged in such service on board the tanker at any time.

*It is recommended that wherever possible the MNTB Stage II (four-day) fire-fighting course should be taken in lieu of the minimum requirement of a two-day course.

7. (a) When an officer or rating satisfies the requirements of either paragraph 4(a) or 4(b) above he should be given a statement to that effect signed by an employer or a master. The statement should include the recipient's full name and discharge book number; specify the type of tanker (i.e. petroleum, liquid chemical or liquefied gas) in which the service was performed; and be endorsed with the company or ship's stamp.

(b) If the statement is presented at a Department of Transport Marine Office together, where appropriate, with evidence of attendance at a fire-fighting course and tanker familiarisation course, a suitable endorsement will be entered in the holder's discharge book.

Officers joining a tanker as Master, Chief Mate, Chief Engineer or Second Engineer

8. The Merchant Shipping (Certification of Deck Officers) Regulations 1980 and the Merchant Shipping (Certification of Marine Engineer Officers) Regulations 1980 require that officers appointed to serve as master, chief mate, chief engineer or second engineer of a tanker should hold a dangerous cargo endorsement to their certificate which is appropriate

to the type of tanker concerned. Details of the requirements to be met for the issue of dangerous cargo endorsements are contained in Merchant Shipping Notice No. M.952.

Officers with direct responsibility for loading, discharging and care in transit of cargo

9. Where ships carry a separate cargo officer designated as such, having direct responsibility for loading, discharging and care in transit of cargo, he also should hold a dangerous cargo endorsement appropriate to the type of tanker concerned.

Merchant Shipping Notice No. M.1075

USE OF CO_2 AND OTHER EXTINGUISHING GASES FOR INERTING PURPOSES

Notice to Shipowners, Masters, Officers and Seamen of Merchant Ships, Shipbuilders and Shiprepairers

1. The purpose of this notice is to draw the attention of Shipowners, Masters and Crews of Merchant Ships, Shipbuilders and Shiprepairers to the dangers of using CO_2 or halogenated hydrocarbon media in fixed fire extinguishing systems on board ships for inerting spaces which contain, or are likely to contain, an explosive mixture of flammable gases or vapours but in which there is no fire.

2. There is evidence to show that when an attempt has been made to inert a space containing an explosive mixture of flammable gases or vapour and air by using the CO_2 fixed gas smothering system to prevent a fire starting within that space, the action has resulted in an explosion caused by electrostatic sparks produced by the CO_2 discharge.

3. Research undertaken on behalf of the Department has indicated that incendive sparks can be also be produced by discharges of Halon 1301 or Halon 1211 from fixed fire extinguishing systems.

4. In general, fixed fire gas or vapour smothering/extinguishing systems using CO_2, Halon 1301 or halon 1211 should not be used to inert any space which contains, or is suspected of containing, a flammable mixture of vapour or gases and air.

5. In their publication, Survey of Fire Appliances, Instructions for the Guidance of Surveyors, the Department already requires notices warning of the danger of using CO_2 for inerting pumprooms or cargo tanks of tankers to be posted at the operating controls. Similar notices should be posted at the controls of systems using Halon 1301 or Halon 1211 for the protection of pumprooms and similar spaces where an accumulation of explosive vapour/air mixtures is a possibility.

6. However, if a fire is known to exist in the protected spaces then the hazard of an explosion does not arise and such systems can be brought into operation with confidence to extinguish the fire.

7. For the reasons given in paragraph 3, fixed systems using Halon 1301 stored in containers within the pumproom and similar spaces are no longer acceptable on new ships because of the possibility of discharge due to the failure of a bursting disc or similar relief arrangements to prevent overpressure in the containers. **NEVER ASSUME THAT THE SHIP IS GAS FREE.**